# Communications in Computer and Information Science 3001

Series Editors

**Rationale**
The CCIS series is devoted to the publication of proceedings of computer science conferences. Its aim is to efficiently disseminate original research results in informatics in printed and electronic form. While the focus is on publication of peer-reviewed full papers presenting mature work, inclusion of reviewed short papers reporting on work in progress is welcome, too. Besides globally relevant meetings with internationally representative program committees guaranteeing a strict peer-reviewing and paper selection process, conferences run by societies or of high regional or national relevance are also considered for publication.

**Topics**
The topical scope of CCIS spans the entire spectrum of informatics ranging from foundational topics in the theory of computing to information and communications science and technology and a broad variety of interdisciplinary application fields.

**Information for Volume Editors and Authors**
Publication in CCIS is free of charge. No royalties are paid, however, we offer registered conference participants temporary free access to the online version of the conference proceedings on SpringerLink (http://link.springer.com) by means of an http referrer from the conference website and/or a number of complimentary printed copies, as specified in the official acceptance email of the event.

CCIS proceedings can be published in time for distribution at conferences or as post-proceedings, and delivered in the form of printed books and/or electronically as USBs and/or e-content licenses for accessing proceedings at SpringerLink. Furthermore, CCIS proceedings are included in the CCIS electronic book series hosted in the SpringerLink digital library at http://link.springer.com/bookseries/7899. Conferences publishing in CCIS are allowed to use our online conference service (Meteor) for managing the whole proceedings lifecycle (from submission and reviewing to preparing for publication) free of charge.

**Publication process**
The language of publication is exclusively English. Authors publishing in CCIS have to sign the Springer CCIS copyright transfer form, however, they are free to use their material published in CCIS for substantially changed, more elaborate subsequent publications elsewhere. For the preparation of the camera-ready papers/files, authors have to strictly adhere to the Springer CCIS Authors' Instructions and are strongly encouraged to use the CCIS LaTeX style files or templates.

**Abstracting/Indexing**
CCIS is abstracted/indexed in DBLP, Google Scholar, EI-Compendex, Mathematical Reviews, SCImago, Scopus. CCIS volumes are also submitted for the inclusion in ISI Proceedings.

**How to start**
To start the evaluation of your proposal for inclusion in the CCIS series, please send an e-mail to ccis@springer.com

Monica Bianchini · Marco Gori · Marco Maggini
Editors

# Artificial Life and Evolutionary Computation

19th Italian Workshop, WIVACE 2025
Siena, Italy, September 3–5, 2025
Proceedings

Springer

*Editors*
Monica Bianchini
University of Siena
Siena, Italy

Marco Gori
University of Siena
Siena, Italy

Marco Maggini
University of Siena
Siena, Italy

ISSN 1865-0929 ISSN 1865-0937 (electronic)
Communications in Computer and Information Science
ISBN 978-3-032-33184-7 ISBN 978-3-032-33185-4 (eBook)
https://doi.org/10.1007/978-3-032-33185-4



This Springer imprint is published by the registered company Springer Nature Switzerland AG
The registered company address is: Gewerbestrasse 11, 6330 Cham, Switzerland

# Preface

This volume of *Communications in Computer and Information Science* contains the proceedings of WIVACE 2025, the XIX Workshop on Artificial Life and Evolutionary Computation. The workshop took place from September 3 to 5, 2025, at the Department of Information Engineering and Mathematics of the University of Siena, Italy, and was organized as a satellite event of the 21st Conference of the Complex Systems Society (CCS 2025). The International Workshop on Artificial Life and Evolutionary Computation aims to bring together researchers from diverse disciplines to present and exchange ideas, results, and perspectives within a multidisciplinary setting. It serves as a forum for discussing emerging research directions and innovative applications in Artificial Life, Evolutionary Computation, and related areas, fostering interaction among different scientific communities. In particular, the workshop promotes collaboration among computer scientists, mathematicians, biologists, psychologists, and cognitive scientists to explore topics such as the modeling and simulation of life mechanisms, evolution and adaptation, and collective and social behaviors, with the goal of advancing technological solutions inspired by biological principles. The workshop was organized over two and a half days and featured single–track sessions including 11 oral presentations and 5 poster presentations. Its format was specifically designed to encourage interaction among doctoral students and early–career researchers, providing a collaborative and constructive environment in which participants could discuss ideas and disseminate their work. WIVACE 2025 also featured two keynote speakers, who opened the morning sessions of the two main workshop days: Stefano Melacci, from the University of Siena, and Diego Garlaschelli, from IMT School for Advanced Studies Lucca (Italy) and Leiden Institute of Physics (The Netherlands). Their keynote talks, respectively titled *Collectionless Artificial Intelligence: Peer–to–Peer Networks of Agents for Decentralized Learning Over Time* and *Multi–Scale Node Embeddings for Graph Modeling and Generation*, provided stimulating perspectives on current research challenges at the intersection of artificial intelligence, complex systems, and network science. Overall, the workshop received 15 submissions, of which 14 were selected for inclusion in this volume following a two–round single–blind review in which each submission received an average of two reviews from members of the Program Committee. We would like to express our sincere gratitude to all the PC members for their valuable contribution to the review and selection process. We also take this opportunity to thank the Local Organizing Committee — Simone Bonechi, Barbara Toniella Corradini, and Marco Stefanelli of the Siena Artificial Intelligence Lab (SAILab) at the University of Siena — for their essential support in organizing the event. Last but not least, we are grateful to the staff of the University of Siena and the host institution for providing the facilities and infrastructure that made the workshop possible.

September 2025

Monica Bianchini
Marco Gori
Marco Maggini

# Organization

## General Chairs

| | |
|---|---|
| Monica Bianchini | University of Siena, Italy |
| Marco Gori | University of Siena, Italy |
| Marco Maggini | University of Siena, Italy |

## Local Organizing Committee

| | |
|---|---|
| Simone Bonechi | University of Siena, Italy |
| Barbara Toniella Corradini | University of Siena, Italy |
| Marco Stefanelli | University of Siena, Italy |

## Program Committee

| | |
|---|---|
| Marco Baioletti | University of Perugia, Italy |
| Stefano Cagnoni | University of Parma, Italy |
| Timoteo Carletti | Univerisity of Namur, Belgium |
| Antonio Chella | University of Palermo, Italy |
| Luca Di Gaspero | University of Udine, Italy |
| Francesco Fontanella | University of Cassino, Italy |
| Mario Giacobini | University of Turin, Italy |
| Alex Graudenzi | University of Milano-Bicocca, Italy |
| Giovanni Iacca | University of Trento, Italy |
| Luca Manzoni | University of Trieste, Italy |
| Giancarlo Mauri | University of Milano-Bicocca, Italy |
| Eric Medvet | University of Trieste, Italy |
| Giorgia Nadizar | University of Trieste, Italy |
| Mario Pavone | University of Catania, Italy |
| Anthony Pipe | University of Bristol, UK |
| Andrea Roli | University of Bologna, Italy |
| Alessandra Scotto di Freca | University of Cassino, Italy |
| Roberto Serra | University of Modena and Reggio Emilia, Italy |
| Pasquale Stano | University of Salento, Italy |
| Thomas Stützle | Université Libre de Bruxelles, Belgium |
| Elio Tuci | University of Namur, Belgium |
| Marco Villani | University of Modena and Reggio Emilia, Italy |
| Mathias Weyland | Zurich University of Applied Sciences, Switzerland |

# Editorial

Artificial Life (ALife) and Complex Systems represent an established interdisciplinary research area dedicated to understanding the fundamental principles underlying life, adaptation, and emergence of new collective behaviors. Combining approaches from computer science, artificial intelligence, biology, physics, mathematics, and engineering, this field investigates how complex phenomena arise from the interaction of simple components, in both natural and artificial environments. In this perspective, artificial systems become experimental laboratories for testing hypotheses about natural life and for designing new forms of intelligent worlds. Complex Systems complements this approach by studying networks of interacting entities whose collective dynamics generate emergent properties that cannot be reduced to the behavior of individual components alone. In recent years, the convergence between Artificial Life, Artificial Intelligence, and Complex Systems has opened new research directions with significant scientific and societal implications. Advances in machine learning, bio-inspired computation, ecological modeling, and data-driven simulations are enabling researchers to tackle pressing challenges related to sustainability, environmental management, healthcare, and the understanding of biological diversity and the origins of life. Within this broad scientific landscape, for about twenty years, WIVACE has offered a valuable platform for interdisciplinary exchange, fostering dialog among researchers working on theoretical foundations, computational methodologies, and real-world applications of ALife and Complex Systems. The contributions collected in these proceedings reflect the diversity, vitality, and growing impact of a field that continues to redefine the boundaries between natural and artificial forms of organization and intelligence. Indeed, the WIVACE 2025 proceedings, collected in this volume, constitute a rich and multidisciplinary set of contributions exploring the evolving landscape of ALife and Complex Systems. The volume reflects the lively scientific dialog that emerged during the workshop, bringing together perspectives from computational modeling, life sciences, environmental studies, and complex adaptive systems. Thematic sessions — *Artificial life for environment management*, *AILife: Artificial Intelligence in Life Sciences*, *Going deep in the origin and diversity of life*, and *ALife in complex environments* — highlight both the breadth and the contemporary relevance of research in the field. Across these domains, contributors investigate how artificial and natural systems interact, adapt, evolve, and generate emergent trends, offering new methodological and conceptual tools for understanding complexity in biological, ecological, and social contexts.

# Artificial Life for Environment Management

This section collects three contributions that explore different yet complementary applications of Complex Systems and Artificial Intelligence methodologies to the analysis and management of dynamic real-world environments. Despite focusing on distinct domains — urban systems, intelligent transportation, and geohydrological risk prediction — they all share a common perspective centered on the study of interconnected networks, adaptive processes, and data-driven decision-making to ensure safety and quality of life across urban, environmental, and infrastructural contexts. In particular, the first contribution, *Venezia, the living city*, by U. Hershberg and J.J. Merelo, investigates urban walkability through the network structure of Venice, highlighting how the historical evolution of pedestrian connections reflects the functionality and sustainability of a city. The second paper, *Multi-Agent Learning for Multi-Intersection Traffic Control with a Swarm of Drones*, by D. Alahvirdi and E. Tuci, applies multi-agent deep reinforcement learning and drone-based sensing to optimize traffic light coordination in complex urban traffic networks, demonstrating how intelligent distributed systems can improve mobility and reduce congestion. Finally, the third work, *Accelerating Landslide Temporal Forecasting through a Re-engineered Software System: $^{GA}$SAKe 2.1*, by A. Sidero *et al.*, introduces a re-engineered hydrological model based on genetic algorithms for predicting landslide activation, emphasizing computational efficiency and the modelling of complex environmental dynamics. Together, these studies demonstrate how data-driven methodologies inspired by adaptation and cooperation are successfully applied to pressing societal and environmental challenges.

# AILife: Artificial Intelligence in Life Sciences

In this section, contributions investigate, across different scientific domains, the capability of computational and AI-based methods to extract meaningful and transferable representations from complex biological or chemical data. Specifically, the collected papers explore how advanced machine learning and signal processing techniques can capture latent structures and semantically relevant patterns even in highly specialized contexts, characterized by strong domain specificity and noisy experimental conditions. In the first paper, *A comparison on Cross-Domain Representation Learning for Skin-Lesion Classification*, by Andreini *et al.*, transfer occurs from natural image domains to dermoscopic imaging for skin-lesion classification, showing that pre-trained vision models, including multimodal architectures, produce robust and highly discriminative features even in a specialized medical context. In the second study, *Comparing General and Domain-Specific Pre-trained Language Models for Natural Product Classification*, by B. T. Corradini *et al.*, the focus shifts to the chemical domain, where general-purpose language models are evaluated on the task of natural product classification using molecular representations, achieving performance comparable to or exceeding domain-specific models, even under input perturbations. Finally, the third contribution extends these ideas toward unconventional biological substrates and bioelectronic computation. In *Electrical Dynamics in Living Scenedesmus acutus*, by S. Del Basso *et al.*, the spontaneous and electrically stimulated activity of the microalga *Scenedesmus acutus* is analyzed through computational techniques. The study identifies characteristic low-frequency electrical patterns whose spectral power increases under external stimulation, while remaining absent or minimal in the surrounding medium alone. Beyond the biological characterization itself, the work highlights how computational models can uncover structured and potentially functional dynamics in living substrates, suggesting the possibility of employing microalgal systems as building blocks for future bioelectronic computing architectures.

# Going Deep in the Origin and Diversity of Life

This section gathers four contributions that, while addressing different scientific questions, share the common thread of the emergence of organization, autonomy, and meaningful behavior in complex systems. Whether in prebiotic chemistries, interactive exploration frameworks, thermodynamic information processing systems, or cellular automata, the central question remains the same: how can simple interacting elements generate coherent, adaptive, and potentially meaningful higher-level behaviors? The contributions collectively provide complementary insights into the mechanisms through which complex systems organize themselves and evolve toward increasingly sophisticated forms of structure and dynamics. In particular, the first study, *Collective Autocatalysis in a Modified Binary Polymer Model*, by S. Carpi *et al.*, investigates the emergence of autocatalytic organization in chemical systems relevant to the origin of life. By introducing structure-dependent catalysis into Kauffman's classical polymer model, the study demonstrates how the topology of catalytic interactions strongly shapes the formation and stability of autocatalytic sets, showing that collective organization emerges not merely from the presence of reactions, but from the structured coordination among interacting components. The second contribution extends the discussion to the thermodynamic foundations of biological organization and information processing. In *Thermodynamics of Biological Switches*, by R. D. Jones *et al.*, a formulation of the First Law of non-equilibrium thermodynamics for biological information processing systems is derived, by partitioning entropy into microscopic and mesoscopic components and by assuming that natural selection promotes efficient information processing and transmission. By connecting thermodynamics, information theory, and biological organization, the study provides a theoretical perspective on how ordered and functionally meaningful structures may emerge and stabilize in living systems far from equilibrium. Instead, *Complex System Exploration with Interactive Human Guidance*, by B. Morel *et al.*, addresses complexity from a methodological perspective. Rather than focusing on a specific system, it studies how humans can efficiently explore vast, highly non-linear parameter spaces to discover meaningful emergent patterns. By integrating explicit user-defined constraints into the exploration process, the proposed framework enables a balance between constrained discovery and global diversity. Finally, the paper *A phenomenological glider? Enactivism and IIT in the Game of Life*, by F. Rodriguez and P. Husbands, using the glider structure in Conway's Game of Life as a minimal autonomous system, investigates whether simple self-organizing dynamics can exhibit forms of intrinsic information integration compatible with the Integrated Information Theory of Consciousness. The findings suggest a deep connection between autonomy, self-organization, and phenomenological properties, supporting the hypothesis that cognition-like dynamics and primitive forms of subjective organization may emerge even in highly minimal systems.

# ALife in Complex Environments

The last section gathers four contributions that investigate how complex systems interact with, adapt to, and regulate dynamic environments across different domains, ranging from socio-technical networks and ecological robotics to online social systems and intelligent sensing technologies. Despite their diversity, the papers share a common conceptual framework: complex environments are active, evolving structures that continuously shape—and are shaped by—the adaptive dynamics of the systems embedded within them. Collectively, the contributions explore how resilience, self-regulation, information flow, and adaptive modeling emerge from the interaction between local processes and large-scale environmental organization. The first contribution investigates resilience in large-scale interconnected systems. In *Characterization of Bow-Tie zones in unknown systems*, by C. Uguzzoni and M. Villani, the influence of the structural organization of directed networks on their response to failures and cascading disruptions is analyzed. Focusing on Bow-Tie representations of complex systems such as ecological or infrastructural networks, the work shows how depth-first snowball sampling can efficiently characterize resilience properties even when the global organization of the system is initially unknown. The study highlights how adaptive exploration strategies can reveal critical structural regions responsible for robustness and vulnerability in complex environments. The second paper approaches complexity from an ecological and techno-philosophical perspective. In *Toward a Gaian Synthetic Approach: Techno-Apoptosis as a Catalyst for Sustainable Robotics*, by L. Damiano and A. Fleres, sustainability is reframed not as an external engineering constraint but as a generative principle for synthetic and robotic systems. Building on the idea of understanding by building, a "Gaian turn" is proposed, where artificial systems, such as robots, actively participate in the self-regulation of their ecological contexts. In this perspective, adaptation to complex environments involves not only persistence and robustness, but also transformation and systemic compatibility within broader ecological dynamics. The third contribution focuses on the dynamics of information exchange and polarization in online social environments during large-scale societal disruptions. In *Impact of the COVID-19 Outbreak on the Italian Twitter Vaccination Debate: A Network-Based Analysis*, by V. Lachi *et al.*, the evolution of the Italian Twitter vaccination debate before and after the COVID-19 outbreak is analyzed through large-scale retweet networks. The study reveals how the pandemic triggered a sharp increase in activity, centralization, and polarization, leading to the emergence of strongly influential hubs and highly differentiated ProVax and NoVax communities. By studying how collective behaviors self-organize under conditions of uncertainty and crisis, the work highlights the role of network structure in shaping information propagation, visibility, and social influence in digitally mediated environments. Finally, the fourth paper addresses adaptive sensing in physically complex environments. In *Numerical Modeling of Surface Reactions in Chemoresistive Gas Sensors: A Computational Approach to Predictive Behavior*, by E. Panzardi *et al.*, numerical modeling techniques are applied to chemoresistive gas sensors based on metal oxides in order to describe and predict their dynamic responses under varying gas compositions and thermal conditions. By integrating surface reaction kinetics with electronic conduction mechanisms, the proposed framework provides interpretable computational models capable of capturing both transient and steady-state sensor behaviors. Beyond their engineering relevance, these models establish a bridge between physics-based representations and data-driven inference, showing how intelligent sensing systems can adaptively interpret complex environmental signals.

The contributions collected in this volume demonstrate the richness and diversity of current research in the field of ALife and Complex Systems, while also revealing a set of common underlying questions and methodological convergences. Across different domains, scales, and perspectives, the proceedings of WIVACE 2025 highlight how complex phenomena emerge from the interaction of adaptive processes, information dynamics, and organizational structures. As the research continues to evolve, the field is progressively moving from the analysis of isolated mechanisms toward an integrated understanding of adaptive, embodied, and environmentally embedded forms of organization. Taken together, the presented contributions show that contemporary research increasingly requires integrative approaches capable of bridging traditionally separated disciplines. We hope that the ideas, methods, and perspectives presented here will stimulate further interdisciplinary collaborations and inspire future developments across the broad landscape of ALife and Complex Systems research.

Monica Bianchini
Marco Maggini

# Contents

# Artificial Life for Environment Management

# Venezia the Living City

Uri Hershberg[1](✉) and Juan Julián Merelo Guervós[2]

[1] Department of Human Biology, School of Life Sciences, and Center for Biophysics and Quantitative Biology, University of Haifa, Haifa, Israel
uri@sci.haifa.ac.il

[2] Department of Computer Engineering, Automatics and Robotics + CITIC, University of Granada, Granada, Spain
jmerelo@ugr.es

**Abstract.** What makes some cities successful and others less so? We propose here that the structure of a city and the way it enables motion through its streets; i.e. its walkabilty, predicts its usability. We further suggest that this can be seen by studying the network structure of a city's streets, since the network reflects the history of their use. As an example of our claim we will study Venice. The city is a good example for our hypothesis as it is both universally walkable and one of the most purposefully sustained cities in existence. To study the network of Venice in a way that reflects the need to adapt it to the motion of people we will single out those elements of the city that are most changeable, that is, those that can be easily added or created by eliminating some barrier. We will study the bridges between islands (*ponti*) and contrast them with the intra island streets (*calli*) that make up most of the pedestrian network of Venice. Even in its uniqueness, looking at Venice will help us develop a methodology for analyzing the current state of a city graph and gather some insights on how it achieves its purpose through historical analysis of its evolution.

**Keywords:** Social network analysis · Complex networks · City design · Venice · Complex network evolution · Urban networks · Betweenness centrality · Complex systems · Adaptation

## 1 Introduction

We propose here a study of cities as living systems. Living systems are always in flux, *panta rei* – everything flows. Much like one can never steps into the same river twice, since the essence of the river is the shifting waters that flow in it. Cities are also made of their moving parts. The streets and neighborhoods, as static as they may seem, are like riverbeds, embodiments of the flows they contain. Cities are constantly being rebuilt, they grow and change and adapt but still retain something of their character.

No city better exemplifies this trait than Venice. The city that has been sinking into the lagoon since its inception and yet appears unchanged. What is


M. Bianchini et al. (Eds.): WIVACE 2025, CCIS 3001, pp. 3–16, 2027.
https://doi.org/10.1007/978-3-032-33185-4_1

the secret of Venice's resilience? Can it be learned by studying the networks on which things flow in Venice? Venice is a medieval city that still functions today in pretty much the same form. Like all cities of its time, and to the extent it can be considered "designed", it was designed for the scale of humans and donkeys [17].

In the 15th century Venice had an advantage over other cities because of its canals. These canals and their uses to transport goods and people at high volume, only came to other cities with the invention of subways and trams. However, in other cities acquiring the capabilities of mass transport came at the price of creating urban sprawl and effectively changing the scale of cities from that of man and walking to that of the car. Not so in Venice where outside the canals it is still the most walkable of cities. In Paris, London, New York and Rome urban planners are trying to create modern walkable sustainable city centers. It appears quite clear that if we wish for our city centers to thrive this is what we need: A return to human scale. The idea of walkability is also at the heart of futuristic projects such as NEOM, also known as the Line - the planned city concept of the crown prince of Saudi Arabia, and in the critical response to the Line, which suggested that a more human scale possibility for a futuristic walkable city would be a circular design [18].

We would propose that Venezia, the medieval city, can teach us how to create viable walkable neighborhoods with mixed usage and mixed demographic communities [5]. To start our research we will look at the network structure of the paths of Venice, including all the bridges, to identify their scale(s) and patterns of connectivity and in that way suggest what could be the best design principles for creating viable walkable cities. We would like to connect the network structures of paths to the way people traverse these paths; to do so we will look at the role of the most changeable components of the Venetian pedestrian networks: The bridges (ponti).

Helpfully, the routes of Venice all receive different names, which are related to their history and function [22]. Many of them are called *calle*[1] and connected across canals by *ponti*[2] (bridges). These last elements are much more easily constructed and removed. We therefore hypothesize that when they are added they represent a selective change to the networks and their connectivity. Indeed, as we will show below, when we consider bridges separately we find that their betweenness distribution has a higher value than the rest of the network and that two stretches of bridges with the highest betweenness may reflect how Venice evolved from a centralized mercantile center to more distributed place to live in, be visited and explored.

To test our hypothesis that the city has adapted and that the main loci of adaptation are the bridges we analyzed the graph representation of the urban structure of the city of Venice trying to answer the following questions:

---

[1] In Italy, the most common name for a street is *strada*; Venice is unique in the way it calls its different ways: *calli* (street) or *campo* (squares).

[2] Please note that we will be using the Italian (and sometimes Venetian) convention for plurals throughout this paper.

- What is the large-scale structure of connections in the city? Are different types of ways providing connectivity in an interchangeable way or not? Specifically do the *ponti* (bridges) between islands represent a specific class of connectivity in the network?
- Does the connectivity provided by the edges reflect the adaptability of the city structure in some way? Can we identify key edges that describe how Venice is connected right now and suggest how it is changed over time?

The rest of the paper is organized as follows: Next we will present the state of the art in studying cities from the point of view of its network structure; Sect. 3 will introduce the dataset used for the study and how it has been obtained and processed. Results will be presented in Sect. 4; we will finally discuss them and show our conclusions in Sect. 5.

## 2 State of the Art

Our study aims to combine understanding from several existing points of view of city design and network analysis. This is a preliminary study on which we hope to first develop a algorithmic methodology and then build a more comprehensive analysis incorporating all these aspects.

To understand the design of cities some have suggested that we must understand how the functions of a city combine [8] the way people move through the city [4,7] and how they observe it; that is, how the iconic parts of the city make those navigating through the city achieve a sense of place [11,19]. In many cases, in order to study each aspect a network representation of the city was made where nodes can be either specific locals connected by edges representing routes [6,19] or they may be more general aspects of the city or zones. In studying Venice, however, we are inspired by a book which created both networks of view and of motion in Venice to show how these combine in the constructed logic of the Venice in particular and as an inspiration for city design in general [20].

Once the city is transformed into a network the tools of complex network design and analysis, which have been also used to study other complex living/changing networks, can be applied [8,14], to measuring the underlying network of Venice. We will apply different levels and types of centrality to study different parts of the city's underlying network [3,4,21] and the way it divides into communities or neighborhoods (in the network analysis sense [15,23]).

Of especial interest to us are those attempts at studying the network not as a static map but as a substrate for motion and changes in connectivity as the dynamics of interaction between components of the system emerge. Such methods have been used before to study protein interaction networks [10] and social network interactions [9] but are equally applicable to the study of motion and interactions in a city [12]. Thus one could use methods of community identification based strictly on the structure of the network [15] or on the motion of agents on the network with differing levels of memory related to past choices on motion [23]. Taking this later approach or other simpler choices to identify types of nodes by their structural stability should improve on other analyses that did not consider that the networks we are studying, even if they describe something of apparent stability like a city, represent only the template of potential interactions.

Thus, we will next use the tools of complex network analysis to understand the current urban structure of *Venezia*, and check how it is shaped by the physical constraints, the perception of the city from a personal point of view, and its evolution by the creation of new edges connecting nodal points in the city.

## 3 Methods

There are two sources of data that can inform about how motion occurs in Venice: We could study the motion of individuals or we can study the network of pathways. Although it would be good to combine both viewpoints, we will as a starting point, present here solely a study of the network of pedestrian pathways that connect the islands of Venice.

OpenStreetMap hosts a frequently updated information on Venice (and most cities in the world). We have used the `osmnx` Python library to download the data using the *placename* "Venezia-Murano-Burano, Veneto Italy"; data was downloaded in August 2024. The preparation of the data included these steps

- Extract, using the `graph_from_place command`, the graph for that specific place, using the `walk` network type, to exclude waterways and focus on the actual *calli* of the city. The *calli* compose a graph which, by nature of its creation, is a complex graph, and one that describes how the city of Venice and its different parts can be reached by foot. The nodes of this graph will be junctions of any two segments of footpaths, including of course squares, while edges will be these footpaths that will include all or part of the *calli*, bridges, and other street under different names. This results in more than 14K edges.
- The original graph is directed, we made it undirected by merging all edges with the same origin and destination in any direction.
- Data in the original graph file is attached to edges in the processed graph, by adding attributes that classify them as bridges (*ponti*) and color them according to their traits.
- Save as a graph to a file. This is an XML file in the GraphML format.
- The directed graph is then converted into an undirected one, choosing randomly the type of one of the two directions as the edge type.

Not all paths are of equal importance even before we know how people use them. Beyond the *calle* (street) that connects squares (*campi* and *campielli*) on a given island one needs *ponti* (bridges) to move between islands. Identifying where bridges are and how they impact the scale and connectivity of the network can also give us an indication of places where the networks changed. While changing the pathways of a *calle* requires the creation or destruction of buildings, a simple bridge will connect two existing islands and can be quite crooked[3] to accommodate existing paths on both sides.

[3] There are the so called *ponti storti*, literally crooked bridges.

Even as all these edges are walkable, OSM makes distinctions among them; these different types ("pedestrian", "footway" or "service") have been eventually found irrelevant for the current study, except it can have additional information such as classifying segments in a bridge as "steps"; this is mainly due to the fact that most edges in the graph are forbidden to vehicles so distinctions between different types are really lost. We have, however, left these distinctions in the dataset so they can be used for future studies with a different focus.

Besides bridges, another type of walkway in Venice, which is more easily added or removed is the *sotoportego*, (literally - under the gate) which can be an underpass or, in many cases, simply a open passageway; tourists tend to avoid them, but locals use them extensively for their navigation of the city. Together with bridges, they are a very common feature of any kind of itinerary in Venice. While 6.94% of all edges are bridges, 7.49% are classified by OSM as *tunnels*, which is the closest form they have.

Personally validating the roles of edges we assigned is beyond the scope of this study. However, the open source process at the heart of OpenStreetMap guarantees that eventually any errors will be corrected by local users.

The processed dataset that is used in this paper will also be made available as open data in a GitHub repository.

## 4 Results

The network measurement we are going to work with is edge betweenness, a concept introduced by Newman and Girvan [15] in the context of their proposed community detection algorithms; in general, it represents how important a specific edge is within the network, that is, how many paths from a random part of the network to another would pass through it. It is related to community structure in the sense that, by cutting those links, we would reveal the existence of those communities; of course, it could also be related to the amount of traffic that every edge would need to handle. But from our point of view, it is more interesting to interpret high edge betweenness as a semi-autonomous decision that tries to (in many cases literally) bridge structural gaps so that newly linked communities can be reached, or can be reached faster. That is why it is interesting to look at the bridges' betweenness and their rank.

The edges with the highest betweenness are listed in Table 1. We use the same name of the edge as shown in OpenStreetMap, which sometimes includes two or even more elements. For instance, the first one, `['Ponte del Sepolcro','Riva degli Schiavoni']`, could have a composite name with *Ponte* (or bridge), and *Riva*, which is essentially a promenade by the edge of one of the big *canali*[4] or by the *Bacino di San Marco*, or "San Marco pool", the big expanse of the lagoon between San Marco and the Lido.

The two groups of edges with the highest betweenness are represented in Fig. 1. It should be noted that all those edges are *contiguous*, so even if they

[4] Please remember that there are only 3 "*canali*" in Venice: Canareggio, the Gran Canal and the Canal della Giudeca; the rest are called *rii* or rivers.

**Table 1.** Top edges, ranked by betweenness, showing the name of the edge and its type

| Edge name | Betweenness | Bridge? |
|---|---|---|
| ['Ponte del Sepolcro', 'Riva degli Schiavoni'] | 3812670 | TRUE |
| Riva degli Schiavoni | 3768968 | FALSE |
| Riva degli Schiavoni | 3670990 | FALSE |
| ["Ponte de la Ca' di Dio", 'Riva degli Schiavoni', "Riva de Ca' di Dio"] | 3642880 | TRUE |
| Riva degli Schiavoni | 3621503 | FALSE |
| Riva de Ca' di Dio | 3584368 | FALSE |
| Riva de Ca' di Dio | 3568553 | FALSE |
| ['Riva San Biasio', "Riva de Ca' di Dio", 'Ponte San Biasio delle Catene'] | 3552743 | TRUE |
| Riva degli Schiavoni | 3552725 | FALSE |
| ['Riva degli Schiavoni', 'Ponte de la Pietà'] | 3535148 | TRUE |
| Riva degli Schiavoni | 3528388 | FALSE |
| ["Ponte dell'Accademia", 'Campiello San Vidal'] | 3053339 | TRUE |
| Campo Santo Stefano | 2995551 | FALSE |
| Campo Santo Stefano | 2989780 | FALSE |
| Riva San Biasio | 2988618 | FALSE |

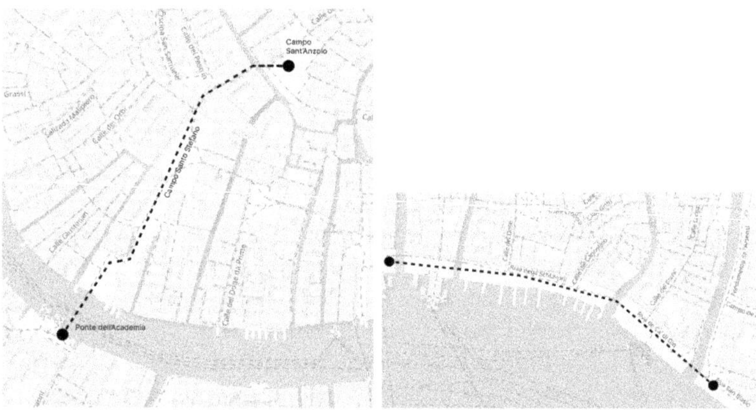


**Fig. 1.** Figure based on annotated screen capture from OpenStreetMap. Black broken line represents edges with the highest betweenness (from the first to the eleventh in Table 1, right, the next cluster of edges, left).

appear as different edges, they are a continuous path that goes from the *Ponte de la Pietà*, through what is one of the most important nodes in the water transportation network, the San Zaccaria vaporetto stop, to right after the *rio del Arsenale*. The Ponte del Sepolcro, with the highest betweenness, is at the left hand side, close to the western end of this path. Besides the fact that the edge with the highest betweenness is a bridge, and 4 of the top-10 are also bridges, another fact that draws our attention is the closeness (but not identification) to the main hub in the city, San Marco, but also the fact that they are all bordering the water, being literally part of the interface between the water transportation system and the on-foot pathways, even if the water edges were not taken into account. This points to an *organic* nature of the growth of the network.

The second block with the highest betweenness goes from the Campo Sant'Anzolo through Campo S. Stefano to the Ponte dell'Accademia[5], also shown in Fig. 1, left. It is not, in this case, bordering a *canal* or a water basin, but going through several *campi* (or squares) ending up in one of four bridges that crosses the Grand Canal[6]. Again, we find three bridges in this path. All in all, 9 bridges are in the top 29 by edge betweenness, that is, 31%, which is beyond the 6.94% that is the actual percentage of edges that qualify as bridges.

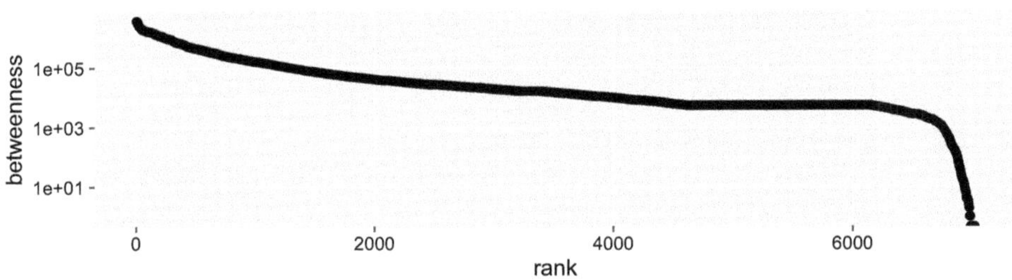


**Fig. 2.** Rank-value plot of edge betweenness

We can get another indication of this by making a log-rank plot of edge betweenness, shown in Fig. 2. This scale shows the well-known broken-stick model, with a slowly descending slope up to a certain point; it goes flat for a few thousand edges and then decreases very fast.

We can look at this from another point of view, shown in Fig. 3, an histogram of edge betweenness values, showing an uneven distribution, where the most abundant value is somewhere around 1000 and 10000, namely, 5519. This number which represents the size (number of nodes) of the Venice network and corresponds to the plateau in the chart in 2. A sampling of the edges included in this set reveals that they are mainly *rami* and *corti*, that is, blind alleys and little squares connected by another alley to a bigger street or *calle*. Along with bridges, another major feature of Venice is that it is limited on all sides by the lagoon. Thus alongside the connections of bridges we had a plenitude of dead ends away from the main thoroughfares. This skews the distribution, but also tells you a lot about the *organic* way this street network was grown to create what is literally a closed fully connected network.

Figure 4 is a violin plot of edge betweenness for bridges and the rest of the edges excluded bridges, which shows that, on average, edge betweenness is much

[5] Except for Riva San Biasio, which is actually contiguous to the previous segment.

[6] Ponte dell'Accademia, Rialto, the Ponte della Costituzione and the *Ponte degli Scalzi* have been built over the Grand Canal.

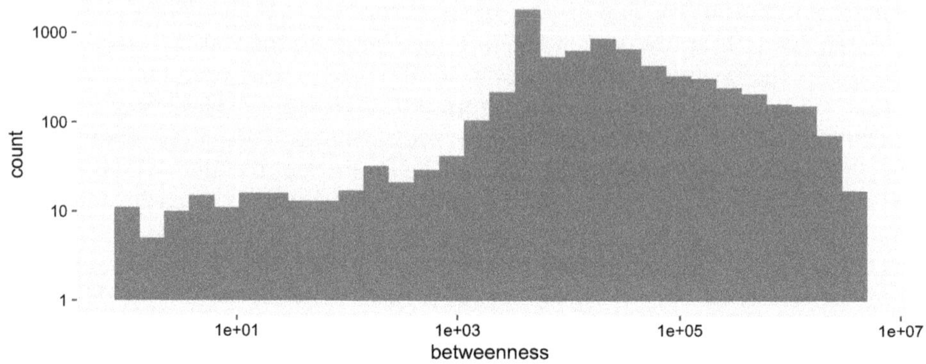


**Fig. 3.** Log-log histogram of edge betweenness value

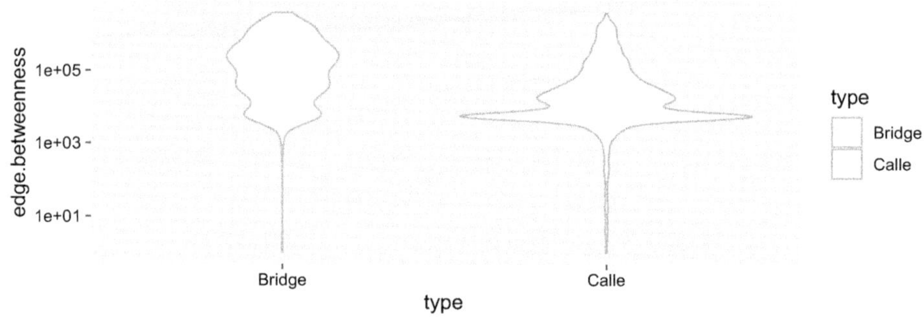


**Fig. 4.** Boxplot of edge betweenness for bridges and calli

higher for the former, and that *calli* or streets do have lots of outliers. The significant difference highlights the fact that bridges have been created in places where a structural gap was perceived, so that they complete the graph in such a way that there are few street (or street segments) that can complete with them, individually or on average.

The median of edge betweenness for bridges is $1.1542023 \times 10^5$ and the rest, which we will generically call *calli*, has $1.3057351 \times 10^4$ has median edge betweenness. Since as we have seen in Fig. 4 the values are not centrally distributed, we have considered the median a better central measure than the average. The difference between medians is significant, median for bridges being 9 times as high as for *calli*. This comparison highlights the fact that bridges are significantly better connectors than the rest of the streets.

Although blind alleys are an important feature in the Venetian urban fabric, its prevalence might obscure larger-scale measures related to way people moves around in the city; in most social networks, anyway, nodes with degree 1 are usually dropped; we therefore next eliminated nodes with a single degree (dead ends) and examined the resulting network.

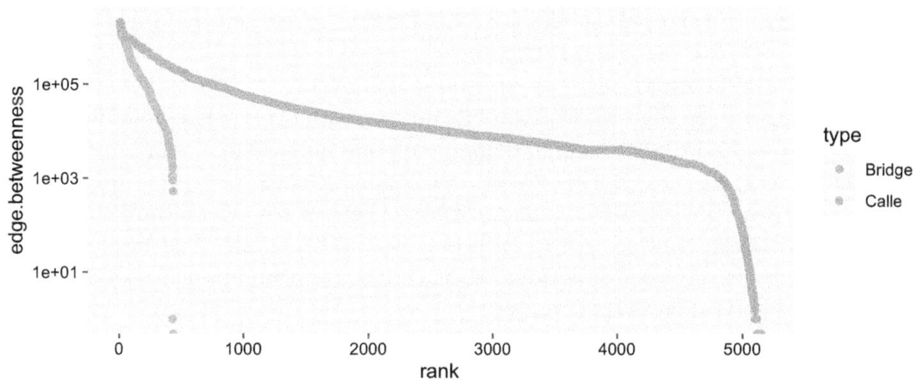


**Fig. 5.** Log-rank plot differentiated by type of path

After pruning *leaves*, we made an edge-betweenness log-rank plot for bridges and the rest of the edges in Fig. 5. The plateau that we saw in Fig. 2 has disappeared in both, although there is a small plateau left in the case of *calli*, probably due to the creation of "new" leaves. The position of the plateau is the same, at the transition between the power law and the descending step. The bridges, however, show a very steep power law (with some outliers), highlighting the fact that every path within a bridge, when eliminated, will isolate a big part of the graph; after pruning, median edge betweenness for bridges is $8.6897001 \times 10^4$ and for calli $1.0325884 \times 10^4$. The median betweenness measure for bridges is still 9 times those for non-bridges, although the multiplier is slightly smaller; when dropping degree-1 nodes, many edges with a low betweenness that connected those nodes have also been dropped, and most of these were *calli*. Still, the structure shows a very solid difference between bridges and other kinds of connectors. This again answers positively research question one: even if we do not consider the blind alleys/degree-1 nodes, bridges provide much higher connectivity to different parts of the city than other types of paths; the only caveat here is that the bridges one would expect to actually be the ones with the highest connectivity, such as the Rialto, are not really on top, although they are in the top-50 s.

Even non-bridges are not an uniform class of way in Venice. Returning to the full graph, including all edges, Table 2 shows how many times specific street names appear in the graph (with those appearing less than 50 times dropped). Every segment identified in the OPEN STREET MAP has one or several street names; for consistencies sake, we have extracted the first word appearing in the first street name of each segment; for instance, `['Campo dei Gesuiti', 'Fondamenta Zen']` would be a *campo*. The first name can be assumed to be random, so this choice does not introduce any kind of bias; this is a needed simplification, however, since analyzing combinations of two types of streets would obscure the analysis of the whole structure. This obviously implies that the table above counts as any type only those segments that have that type in their single street name, or as the type of the first street name in a number of them.

Bearing that in mind, *calle* is the most common, but *fondamenta*, which border canals and the lagoon, also appear quite frequently; other denominations correspond to squares, with *campo* being a big one and *corte* a small one. While *rio* is, effectively, a canal, in this case it corresponds to *rio terà*, or buried canal, streets created by paving over canals, of which there is a good amount in Venice, most of which were created in the 19th century and onward. The name, as can be seen, represent different functions, so it would be interesting if that function is reflected in their edge betweenness.

**Table 2.** Types of street found in the graph (more than 50 times) and their frequency and median edge betweenness, sorted by the latter.

| Segment name | Occurrences | Median betweenness |
|---|---|---|
| Riva | 67 | 311963.53 |
| Ponte | 101 | 100846.11 |
| Salizada | 111 | 43713.46 |
| Fondamenta | 558 | 40509.26 |
| Rio | 184 | 32249.50 |
| Viale | 99 | 23355.58 |
| Calle | 1671 | 16551.00 |
| Campo | 448 | 15435.03 |
| Piazza | 59 | 14060.82 |
| Campiello | 96 | 11864.31 |
| Piazzale | 96 | 11009.79 |
| | 2653 | 9294.76 |
| Corte | 196 | 5519.00 |
| Ramo | 262 | 5519.00 |
| Sotoportego | 94 | 5519.00 |

The edge betweenness for edges with a certain name are shown in the same Table 2. We have already seen (Fig. 1 and Table 1) that the *rive*, together with the bridges that join them, are the segment with the highest betweenness. As indicated, these border the Gran Canal as well as the parts of the city that border the water. In most cases, and despite the image of Venice, there is a path with a varying width between water and the houses; as a matter of fact, *fondamente* is the other kind of thoroughfare that shares this feature. *Ponti*, on the other hand, go over water, and *rii*, which are fifth in the ranking, are paved canals, so we find that 4 out of the top 5 paths in terms of median betweenness are at the same time tangent with water and also consequence of a human intervention at a point in time that tries to avoid structural gaps by creating new edges. The city itself flows over, by the side, and around water, quite literally.

The last type of way in the top 5 in terms of betweenness, which is not related to water, (to the extent that anything in Venice can be) are *salizade*, which means *paved* (street or lane). These streets were not opened, but certainly chosen to be paved over others, that were simply open ground initially; these were [2] the main streets in every parish, paved with bricks in the 13th century. The fact they still maintain a high betweenness is witness to the self-organizing manner that has informed decisions in Venice from the beginning, but also to the way the city has evolved in such a way that most thoroughfares maintain their connective power eight centuries later.

The rest of the ranking is occupied by different kind of streets, with *viale* on top; all these are concentrated in the Castello area, most of them in a part reclaimed from the lagoon in the 1920s. The bottom of the ranking is occupied by no-name segments, *corti*, that is, closed squares, *rami* or alleys, and finally, *sotoporteghi*. These have the same median value because they are mainly dead alleys or squares with a single exit. This edge betweenness value is related to the number of nodes in the network; there is a single path that connects them to the single node at one of their ends. Median value do not fully reflect the variability and diversity of these three kinds of nodes, but further investigation is beyond the scope of this paper.

This answers our second research question: the type of ways which have the highest betweenness are those that can be added in a relatively easy way, by paving over *rii* or bridging over water *ponti* or providing a water-side platform (*fondamente*) for houses lining the water. Since these new edges close structural gaps, they immediately become high-betweenness pathways, reflecting the adaptability of the city structure and making it really a *living* city.

## 5 Discussion and Conclusions

All in all we answered both of our research questions about Venice, characterizing the structure of connections in the city and the special role of bridges in connecting the disparate parts of Venice into a whole. We have further shown that by contrasting the betweenness of different types of edges we can identify meaningful differences in connectivity. Finally, we identified two major connection of points of the Venice network. Interestingly these do not include the two main focus points of Medieval Venice - the Rialto and Piazza San Marco, which are still identifiable by other network analysis means [19,20]. The two foci of betweenness we identified are also not key tourist destinations. Potentially this reflects more recent changes and adaptations of Venice. The construction of the Academia bridge, connecting Dorsoduro with its many academic institutions and student dorms and the expansion of residential neighborhoods into Castello, in areas that before were the boat yards adjacent or related to the Arsenale. Areas that now encompass also the Giardini that host what is now the second key attraction of Venice from May to November, the Art and Architecture Biennale, which started in 1895. A very short while ago in the history of la Serenissima.

The distribution of edge betweenness we observed suggest a network with complex, scale free statistics that is limited by the edge constraints and size of the network [1]. Venice is not infinite: It is in-fact encapsulated by the lagoon. The dead ends and closes *campi* of each island thus influence the distribution of betweenness, indicating for us the size of the network that is Venice. However, the through network of bridges (and calle) exhibit an interconnectedness that allows for the flow and interaction of people across all scales of the city. This kind of structure is found in many places in biology where interaction of cells are important and must be condensed into local tissues; for instance in our lymph nodes, where immune responses are initiated, immune cells move more or less single file through a intricate maze of lymphatics interacting rapidly with a wide array of different cells until they meet a partner immune cell which co-responds to the same pathogenic disease, ensuring response to pathogens and lack of spurious responses to changes in the body. This kind of structure is no less important at the scale of a city, as it enhances the ability to meet people you know, and also new people you may need to know, without getting stuck in one local or limited set of interactions. Better than a line or a circle [18] to enhance human interaction one needs a friendly maze, which is exactly the kind of structure a polity like the republic of Venice needed to ensure a reasonable, and reasonably stable, government [13].

As future lines of work, it would be interesting to analyze the same graph using a *dual* representation, which swaps nodes for edges and vice versa [16]. The fact that the edges with the highest edge betweenness are in fact contiguous segments leads us to think that this approach would help us better identify unique paths that play an important role within the urban fabric. It would also be helpful to perform, to the extent that it is possible, a temporal analysis by dating all edges that can be identified and then proceeding to show how overall connectivity changes; even a simple procedure of elimination of high-betweenness edges would help in that sense, and would help us to identify the structural holes that these edges cover. Finally, as we mentioned at the start, it would be good to combine our analysis of the network of paths in Venice with the analysis of the actual patterns of motion of people on the network.

However, even if only taking into account our simple first step of analyzing the betweenness of edges of different types, we have shown here that Venice can indeed be considered a living adaptive system. Living things maintain themselves. However, this type fo stability is neither static nor the acquisition of some chemical steady state. It is not even the motion between several constant attractors or the classical idea of homeostasis, the active return to a constant state. Instead what we describe here is an emergent stability that depends on the ability to adapt and change the living system's component interactions. Venice remains Venice because of the ability of the network that constructs it to change.

**Acknowledgements.** Work supported by the Ministerio español de Economía y Competitividad (Spanish Ministry of Competitivity and Economy) under project PID2023-147409NB-C21 and Israel Science Foundation (ISF) grant 1327/22.

## References

1. Craig, A., Yücel, M., Muchnik, L., Hershberg, U.: Impact of finite size effect on applicability of generalized fractal and spectral dimensions to biological networks. Chaos, Solitons Fractals **164**, 112707 (2022). https://doi.org/10.1016/j.chaos.2022.112707
2. Douglas, H.A.: Venice on Foot. Read Books Ltd (2012)
3. Girvan, M., Newman, M.E.J.: Community structure in social and biological networks. Proc. Nat. Acad. Sci. United States Am. **99**(12), 7821–7826 (2002). http://eutils.ncbi.nlm.nih.gov/entrez/eutils/elink.fcgi?dbfrom=pubmed&id=12060727&retmode=ref&cmd=prlinks
4. Hanna, S.: The natural scale of urban networks: intelligibility is not scale-free. In: Conference Name: The 14th International Space Syntax Symposium (SSS14) Meeting Name: The 14th International Space Syntax Symposium (SSS14), Space Syntax Network, vol. 14. Cyprus Publisher, Nicosia (2024). https://cyprusconferences.org/14sss/
5. Jacobs, J.: The Death and Life of Great American Cities. Random House (1961)
6. Lagarias, A., Prastacos, P.: Fractal dimension of European cities: a comparison of the patterns of built-up areas in the urban core and the peri-urban ring. Cybergeo: Eur. J. Geogr. (2021). https://doi.org/10.4000/cybergeo.37243
7. Li, C., Psarra, S., Hanna, S.: Traversability in spatial configuration: some theoretical and practical aspects. In: Conference Name: The 14th International Space Syntax Symposium (SSS14) Meeting Name: The 14th International Space Syntax Symposium (SSS14), Space Syntax Network, vol. 14. Cyprus Publisher, Nicosia (2024). https://cyprusconferences.org/14sss/
8. Liu, X., Li, X., Shi, X., Zhang, X., Chen, Y.: Simulating land-use dynamics under planning policies by integrating artificial immune systems with cellular automata. Int. J. Geogr. Inf. Sci. **24**(5), 783–802 (2010). https://doi.org/10.1080/13658810903270551
9. Louzoun, Y., Muchnik, L., Solomon, S.: Copying nodes versus editing links: the source of the difference between genetic regulatory networks and the WWW. Bioinformatics **22**(5), 581–588 (2006). https://doi.org/10.1093/bioinformatics/btk030
10. Luo, F., Yang, Y., Chen, C.F., Chang, R., Zhou, J., Scheuermann, R.H.: Modular organization of protein interaction networks. Bioinformatics **23**(2), 207–214 (2007). https://doi.org/10.1093/bioinformatics/btl562
11. Lynch, K.: City Sense and City Design: Writings and Projects of Kevin Lynch. MIT press (1995)
12. Marshall, W.E., Garrick, N.W.: Effect of street network design on walking and biking. Transp. Res. Rec.: J. Transp. Res. Board **2198**(-1), 103–115 (2011). http://trb.metapress.com/openurl.asp?genre=article&id=d
13. Molinari, M.C.: How the Republic of Venice chose its doge: lot-based elections and supermajority rule. Const. Polit. Econ. **31**(2), 169–187 (2020)
14. Nes, A.V., Yamu, C.: Space Syntax. The Virtual and the Real in Planning and Urban Design, p. 136 (2017). https://www.academia.edu/44628369/Space_syntax_A_method_to_measure_urban_space_related_to_social_economic_and_cognitive_factors
15. Newman, M.E.J., Girvan, M.: Finding and evaluating community structure in networks. Phys. Rev. E **69**(2), 026113 (2004). https://doi.org/10.1103/PhysRevE.69.026113

16. Porta, S., Crucitti, P., Latora, V.: The network analysis of urban streets: a dual approach. XXPhys. A **369**(2), 853–866 (2006)
17. Pounds, N.: The Medieval City. Bloomsbury Publishing, USA (2005)
18. Prieto-Curiel, R., Kondor, D.: Arguments for building the circle and not the line in Saudi Arabia. NPJ Urban Sustain. **3**(1), 35 (2023)
19. Psarra, S.: The role of spatial networks in the historic urban landscape: learning from Venice in the fifteenth and sixteenth centuries. Hist. Environ.: Policy Pract. **9**(3–4), 249–273 (2018). https://doi.org/10.1080/17567505.2018.1516380
20. Psarra, S.: The Venice Variations. UCL Press (2018). https://www.uclpress.co.uk/products/107619
21. Rhrissorrakrai, K., Gunsalus, K.C.: MINE: module identification in networks. BMC Bioinform. **12**(1), 4581 (2011). https://doi.org/10.1186/1471-2105-12-192
22. Tassini, G.: Curiosità veneziane ovvero origini delle denominazioni stradali di Venezia. Fontana, Melchiorre (1882)
23. Yücel, M., Muchnik, L., Hershberg, U.: Detection of network communities with memory-biased random walk algorithms. J. Complex Netw., cnw007 (2016). http://comnet.oxfordjournals.org/content/early/2016/04/22/comnet.cnw007.abstract

# Multi-agent Learning for Multi-intersection Traffic Control with a Swarm of Drones

Davoud Alahvirdi(✉) and Elio Tuci

Robotics Lab, Computer Faculty of Namur University, 5000 Namur, Belgium
{davoud.alahvirdi,elio.tuci}@unamur.be
https://www.naxys.be/robotics

**Abstract.** This paper presents a simulation framework for optimising traffic light signal durations at nine interconnected intersections using multi-agent deep reinforcement learning and swarm of drones. In the first phase, Vissim traffic simulation data is used to online train the deep reinforcement learning, aiming to minimise vehicle congestion on interconnecting roads between intersections. Each intersection employs a deep Q-network method, incorporating state, action, and reward data from neighbouring agents into its loss function during learning. In a subsequent phase, a swarm of drones is deployed to count vehicles on the roads based on predefined trajectories. At specific locations called meeting points, drones transfer collected data to traffic lights and, in return, receive information about other routes monitored by other drones. Each intersection is modelled as a trained agent, while the drone data act as the input states for the network to select the best actions to minimise costs. The simulation results demonstrate the impact of the proposed method in the multi-intersection to reduce traffic congestion at interlinked roads compared with fixed green traffic light timing.

**Keywords:** Traffic Monitoring and Management · Multi-Intersection Control · Vissim Software · Swarm of Drone · Deep Q-Network

## 1 Introduction

Traffic congestion is a growing challenge in urban environments, leading to increase travel time, fuel consumption, and environmental pollution. Also, the limited capacity of road networks compared to traffic demand creates major challenges for effective traffic management, particularly in city centres with interconnected intersections. Fixed traffic signal controllers and stationary sensor-based traffic management systems can be used to manage traffic congestion and to alleviate its consequences. However, due to their static nature, they suffer from: i) reduced flexibility in tracking the dynamic nature of traffic, which does not always congest in the same place; ii) relatively high costs of maintenance and redeployability in different parts of the road network. In recent years, the problem


M. Bianchini et al. (Eds.): WIVACE 2025, CCIS 3001, pp. 17–30, 2027.
https://doi.org/10.1007/978-3-032-33185-4_2

of multi-intersection traffic signal control has attracted significant interest, with multiple studies in the literature focusing on designing coordinated strategies for adjacent intersections to enhance overall performance using techniques such as genetic algorithm (GA) [8], vehicle-to-infrastructure technology (V2I) [4], deep learning (DL), or deep reinforcement learning (DRL) [7]. Multi-agent deep reinforcement learning algorithm (MARL) seems particularly promising solution to develop adaptive traffic signal control (ATSC) since it has achieved better performances than alternative methods [16]. In this algorithm, each agent is treated as an independent learning entity that receives road network state information from distributed sensors and selects actions based on its policy. The agent's actions affect future road network states, which are used to reward or penalise current agent policy. In this study, Q-learning is employed to address the ATSC problem across multiple intersections, enabling agents to select optimal actions based on the highest Q-values [17].

In transportation networks, traffic management and monitoring play a key role in improving overall efficiency. Unmanned aerial vehicles (UAVs), commonly known as drones, are widely used for traffic surveillance and data collection. Their mobility, adaptability, and broader coverage make them more effective than traditional fixed cameras. Deploying a swarm of drones enables faster coverage of extensive areas compared to fixed infrastructures. Data about current state of road network gathered by drones can be transferred to traffic station, which can use it for a variety of purposes such as: i) surveillance and monitoring, ii) recognition of traffic violations, iii) management of traffic congestion [6], iv) signal optimization and tracking of vehicles' trajectories [12]. Compared to fixed monitoring systems based on sensors networks, drone-based system can monitor traffic in larger areas that can change based on current traffic conditions.

This paper is structured as follows. In Sect. 2, we review relevant literature with a particular attention to research works based on the use of drones to monitor and manage traffic with DQN. The main methodology of the simulation is explained in Sect. 3. Sections 4 details the traffic model implemented using Vissim software and the structure of the urban road network. The DQN method for traffic signals control is presented in Sect. 5. Section 6 defines the trajectory planning for drones movement. Simulation results are shown in Sect. 7. Finally, a brief conclusion is given in Sect. 8.

## 2 Related Work

Recently, several approaches based on RL have been developed to improve the design of ATSC algorithms. In [7], the authors develop an ATSC system based on deep Q-network (DQN) for multi-intersection networks to improve waiting time, queue length, and to reduce emissions of $CO_2$. In [15], by using MARL, each intersection agent independently learns from camera data and makes decisions based on its own observations and those of neighbouring agents, with the goal of maximizing rewards in the context of a multi-intersection traffic control optimization problem. In [5], the authors propose a cooperative deep Q-network with Q-value transfer (QT-CDQN) for adaptive multi-intersection signal control to minimize average queue length and average waiting time.

As mentioned above, drones play a key roles in smart cities to monitor and collect data about traffic, crowds, and environmental conditions. The data collected by drones can help to improve the management of public services, urban policies, sustainability, resource and waste management. In [13], a swarm of drones is deployed to monitor and gather data to create a general map containing data on traffic, pedestrians, climate, and pollution. Drones are able to connect to traffic unit to transfer data as in [11]. In [10], a cloud-based approach is exploited in a ground station unit for managing drones in a smart city. In [18], by using drones-collected information, the authors design an intelligent traffic control system based on a fuzzy logic controller to reduce average delay time.

This paper aims at further developing a drone-based technology for traffic monitoring and management, by looking at a multi-intersection traffic control scenario. In particular, we develop and test a MARL system, in which agents, receiving traffic data from drones, are trained via the DQN method to determine optimal intersection control actions. We also evaluate the efficiency of a swarm of drones in handling intersection interactions compared to traditional fixed timing traffic light signals. A swarm of four drones continuously monitor road traffic and transfer traffic data to traffic lights governing the flux of traffic at main road intersections. Due to the challenges in acquiring real traffic data for the selected scenario, we model traffic flow using the Vissim simulator [14]. Vissim is a microscopic traffic simulation software developed by the PTV (Planung Transport Verkehr) Group, widely used by transportation engineers, planners, and researchers for analyzing traffic patterns, driver behavior, and infrastructure changes in both urban and highway networks [9]. In our model, each agent, corresponding to a single intersection, is equipped with an individual DQN algorithm, trained using own and other agents data to minimize the number of remaining vehicles on the roads. For evaluation of the trained network, the number of vehicles on the roads network are transferred from the Vissim software to a cellular model in Python. We use a data-driven approach with the swarm of drones to monitor the data. To solve communication problems (e.g., limited bandwidth or data security issues) between drones and traffic lights, specific meeting points are considered to share the traffic information, as in [19]. Traffic lights receive the data from the drones and adapt the green traffic lights duration to minimize congestion.

## 3 Methodology

This Section describes the methods used to develop the simulation modelling of an urban scenario with a swarm of drones carrying out the traffic monitoring and management task. Our study aims to train a multi DQN using real-time traffic data in a simulated $3 \times 3$ intersections aims to reduce the number of remaining vehicles in the central interlink connections by swarm of drones. In the remaining of this section, we illustrate the DQN for optimization of traffic

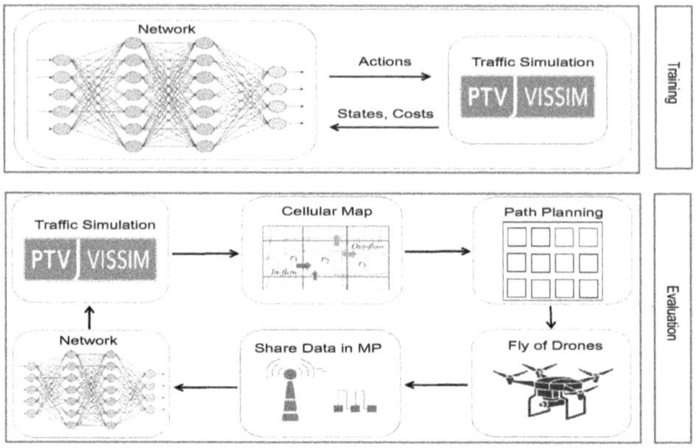


**Fig. 1.** Total steps of the simulation.

lights signal' duration, the Vissim software for realistic traffic modelling, and the swarm of drone for monitoring and collecting traffic data.

The simulation undergoes two phases: the training and the evaluation phases, as shown in Fig. 1. The training phase consists of a geographically accurate representation of a real signalized multi-intersection layout simulated using Vissim. The traffic flow is executed by the Vissim application, which enhances accuracy by considering some realistic features such as vehicle acceleration, deceleration, lane changing, and the presence of various types of vehicles. The main goal of the DQN is to generate actions that efficiently reduce congestion on interlinked roads at the centre of the intersections and limit the impact of traffic flow from neighbouring intersections. During training, actions from the DQN are applied to the traffic lights in Vissim, and traffic states along with associated costs are retrieved to the network.

During the evaluation, the number of vehicles in each subcells of a cellular map representing the road network, are continuously updated based on the traffic modelled in Vissim simulator. Four drones, flying independently at different speed on predefined paths along the road network, monitor the status of the cellular map. To solve communication problems between the drones and the traffic lights, fixed meeting points have been assigned where drones transmit recent collected data to the traffic lights and recieve traffic-related data about regions outside their paths. Thanks to drones, the traffic lights gain knowledge of current traffic in a large portion of the road network and use this knowledge to adjust the green light duration in order to the system goal. Finally, the optimal green time durations are applied to the Vissim software.

## 4 Traffic Model

Vissim is used to simulate traffic dynamics, to generate real-time data for training DQN and to evaluate the traffic light signalling system with respect to changes

**Fig. 2.** Simulated network

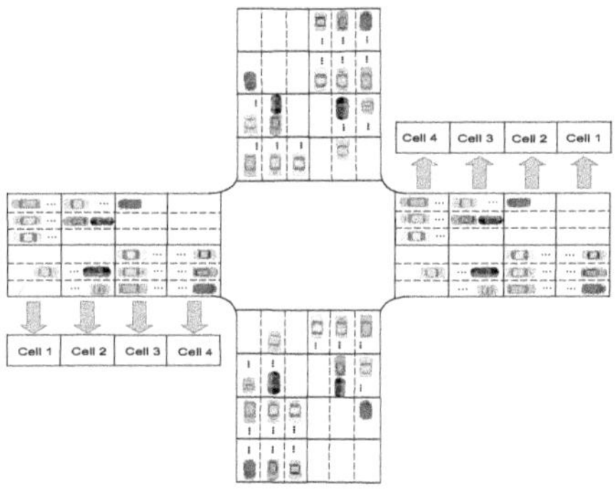


**Fig. 3.** Cellular map

in traffic congestion. Vissim simulates traffic at a microscopic level, representing individual vehicles by specific characteristics, including vehicle interactions, car-following, lane-changing, and mixes different types of vehicles [1]. Figure 2 illustrates the city network as implemented in Vissim, comprising nine intersections. Vissim plays a crucial role in generating data in terms of number of vehicles for training the DQN, which, in turn, receives actions and sends back both states and the associated action costs. It is not feasible to implement drone monitoring of the road network directly in Vissim. A cellular map recreates a discrete representation of Vissim traffic by distributing vehicles in different subcells of the map. The drones monitor the map cell by cell based on predefined map.

Since Vissim roads have a total length of 120 m, the cellular map divides all streets in all directions into four subcells, each measuring 30 m. In each road, the width of the forward-right (hereafter, $fr$) lanes is twice as wide as the left (hereafter, $l$) lanes. The evaluation of the complete traffic management and monitoring system via Vissim is carried out by comparing a fixed traffic light and the adaptive traffic light timing driven by DQN. During both training and evaluation, a fixed number of vehicles enter each $fr$ and $l$ lane to run traffic flow.

### 4.1 Urban Road Network

The cellular map for each intersection consists of four legs: west, east, north, and south as shown in Fig. 3. Each leg of the intersection is structured with two entry and two exit points. One entry lane is designated for vehicles travelling forward or turning right, while the other is assigned to vehicles turning left. Similarly, the first exit point is considered for receiving vehicles travelling forward and for those leaving the intersection with a right turn, and the other exit point is for vehicles leaving the intersection with a left turn. In terms of notation, each intersection is referred to as $R_{\{i,j,k\}}$, with $i$ the number of the intersection, $j = \{IN_{fr}, IN_l, OUT_{fr}, OUT_l\}$ the lanes of each leg, $IN_{fr}$ to indicate the lanes for vehicles entering the intersection to move forward and turn right, $IN_l$ is assigned to the lanes for vehicles entering the intersection to turn left, $OUT_{fr}$ denotes the lanes for vehicles leaving the intersection with forward movement and right turn, and $OUT_l$ represents the lanes used for the vehicles leaving the intersection with left turn; $k = \{W, E, N, S\}$ refers to the four legs direction of the intersection using the cardinal points.

## 5 Proposed Method

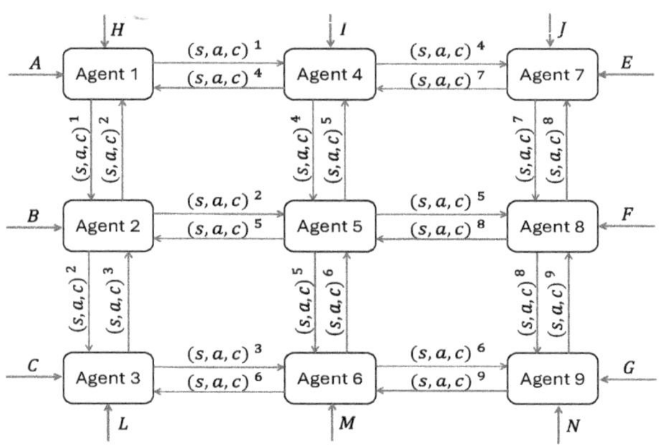


**Fig. 4.** DQN Definition-$(s, a, c)$ refers to state, action, and cost of agent $i$.

Each intersection operates as an independent agent with its own traffic signal controller equipped with DQN, while the outcomes of their actions are interconnected with other agents to minimize the traffic jam at intersection number

5. Hence, the state, action and cost information of adjacent agents are transferred in this model as shown in Fig. 4. By using a multi-agent Markov Decision Process (MMDP) framework, the problem can be described as a 6-tuple $\langle AG, S, A, P, C, \gamma \rangle$, where $AG$ is the set of agents, $S$ is the set of states, $A$ is the set of joint actions, $P$ is the state transition probability matrix, $C$ is the set of costs, and $\gamma \in [0, 1]$ is a discount factor used to measure the importance of future [7]. In this framework, at each time step $t$, the agent $AG_i$ observes the state $S_{it}$ and takes an action $A_{it}$ according to a defined policy. In addition, the selected action leads to a change in the environment from the current state $S_{it}$ to a new state $S_{i(t+1)}$. Consequently, the agent observes $S_{i(t+1)}$ and receives a cost $C_{i(t+1)}$ based on the defined policy. The policy $\pi$ is formulated as:

$$\pi : A_i S_{it} \tag{1}$$

$$\pi(a, s) = P(a_t = aS_{it}) \tag{2}$$

where $P$ represents the probability of transition from state$(S_t = s)$ to state $S_{t+1}$ under action $a$. The estimated Q-values of actions under $s_t^i$ are represented as the output vector of the network. Accordingly, the agent determines the optimal phase for the intersection by selecting the action with the highest Q-value to regulate traffic flow. Using the following formula, the Q-function predicts future costs based on state and action:

$$Q_{t+1}(s_t, a_t) = Q_t(s_t, a_t) + \alpha[C_t + \gamma max_a Q_t(s_{t+1}, a') - Q_t(s_t, a_t)] \tag{3}$$

$Q_{t+1}(s_t, a_t)$ is the updated Q-function value after the learning step $s_t$, while $s_{t+1}$ is the state achieved after executing $a_{t+1}$ and $\alpha \in [0, 1]$ is the learning rate.

### 5.1 Cooperation of Adjacent Agents

The selection of action for an agent is influenced by its own state, action, and cost values alongside with the values of its adjacent agents. In the training phase, we access online data from all agents, but in evaluation mode, drones exchange state, action, and cost values among agents. So, we can rewrite Eq. 3 as:

$$\begin{aligned} Q_{t+1}^i(s_t^i, a_t^i) = & Q_t^i(s_t^i, a_t^i : \theta_i) + \alpha(t)[r_t^i + \\ & \gamma max_{a'} Q_t^i(s_{t+1}^i, a' : \theta_i^*) - Q_t^i(s_t^i, a_t^i : \theta_i)] + \sum_{j \in N} r_{t-1}^j \end{aligned} \tag{4}$$

where $N$ is the number of adjacent intersections and $\theta_i$, $\theta_i^*$ are the evaluation and target parameters of networks, respectively. By transmitting the optimal Q-function values of neighbouring agents to the loss function of the Q-network during policy learning, the system can efficiently coordinate traffic signal control at multiple intersections to achieve the main goal. The loss function of each agent is defined as follows:

$$MSE(\theta_i) = \frac{1}{m} \sum_{t=1}^{m} [[C_t^i + \gamma max_{a'} Q_t^i(s_{t+1}^i, a' : \theta_i^*) + \sum_{j \in n_a} C_{t-1}^j] - Q_t^i(s_t^i, a' : \theta_i)]^2 \tag{5}$$

where $m$ is batch size, $max_{a'} Q_t^i(s_{t+1}^i, a' : \theta_i^*)$ is the optimal target value for all actions under the state $s_{t+1}^i$. $Q_t^i(s_t^i, a_t^i : \theta_i)$ is the output of evaluation network. Thus, the target value noted $y_t^i$ is calculated by the following equation:

$$y_t^i = C_t^i + \gamma max_{a'} Q_t^i(s_{t+1}^i, a' : \theta_i^*) \tag{6}$$

More detailed information about structure of the DQN consists of hidden layers, number of neurons, and activation function are described in [7].

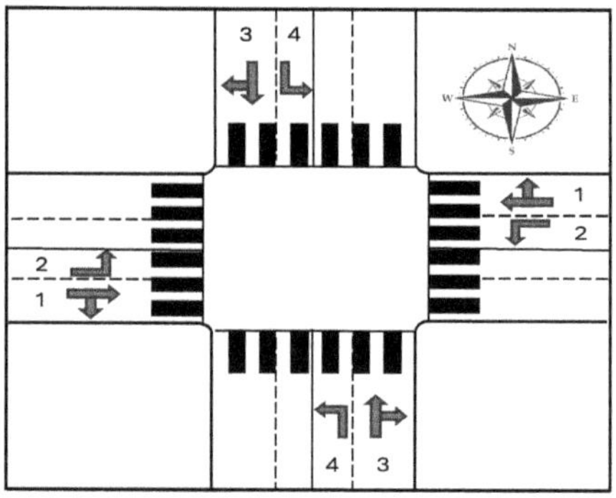

Fig. 5. Traffic light phases

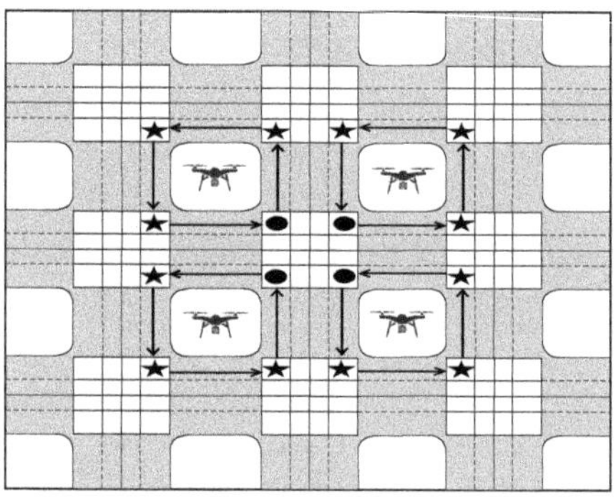

Fig. 6. Path planning of drones

## 5.2 Cost Design

Costs provide feedback from the Vissim environment to evaluate the effectiveness of selected actions. In our configuration, we define the cost as the remaining number of vehicles at the end of each assigned green traffic light duration. For instance, the costs for $fr$ and $l$ lanes in the west/east directions are calculated at the beginning of traffic light phase 2 and 3, respectively. Similarly, the costs for the $fr$ and $l$ lanes in the north/south directions are calculated at the beginning of traffic light phases 4 and 1, respectively. The cost for intersection 5 is calculated as follows:

$$C_5 = a_5(veh_5) \tag{7}$$

The cost for other intersections are formulated as:

$$C_i = a_{i5}(veh_5) + a_i(veh_i) \tag{8}$$

where $a_{i5}$ characterizes the influence of other intersections on the dynamics of intersection 5. Therefore, $a_{15}$, $a_{35}$, $a_{75}$, and $a_{95}$ are all set to zero. Taking the correct action for all agents intuitively leads to a reduction in the remaining number of vehicles on the interlink roads at intersection 5. When the cost is reduced to its minimum, this is an evidence that the model has been learned.

### 5.3 State Space

The goal of designing state spaces is to provide a structured, practical, and simplified representation of agent observations. In the evaluation phase, the state representation is determined by the total vehicle count recorded by drones for each leg of an agent, as described in equation (9). During training, real-time data from Vissim is directly fed into the DQN.

$$S_i = \{S_{iwest}, S_{ieast}, S_{inorth}, S_{isouth}\} \tag{9}$$

The state $S_{iwest}$ represents the total vehicle count in the *west* direction in agent $i$, obtained by summing the vehicles in the $fr$ and $l$ lanes.

### 5.4 Action Space

The set of possible actions for each agent is defined through four phases, as shown in Fig. 5. Each phase consists of three signal lights (red, yellow, and green). We denote signal controls as $S_{\{i,j,k,c\}}$, where $i = \{1, 2, ..., 8, 9\}$ is the number of agents, $j = \{fr, l\}$ represents the lane, $k = \{W, E, N, S\}$ corresponds to the cardinal directions at the intersection, and $c = \{red, yellow, green\}$ indicates the traffic light colors. A detailed description of the phases is provided as follows.

- Phase 1: In this phase, vehicles travel straight from the $W$ to the $E$ leg and vice versa. Since the same traffic signal applies to both straight-through and right-turn movements, vehicles move from the $W$ to the $S$ leg and from the $E$ to the $N$ leg. As a result, the signal for legs $W$ and $E$ are $S_{\{i,fr,W,green\}}$ and $S_{\{i,fr,E,green\}}$ in $fr$ lanes, $S_{\{i,l,W,red\}}$ and $S_{\{i,l,E,red\}}$ in $l$-lanes. All other phases for the $N$ and $S$ legs are in a red light situation: $S_{\{i,fr,N,red\}}$ and $S_{\{i,l,N,red\}}$ for the $N$ leg, $S_{\{i,fr,S,red\}}$ and $S_{\{i,l,S,red\}}$ for the $S$ leg.

- Phase 2: In this phase, vehicles turn left from the $W$ to the $N$ leg and from the $E$ to the $S$ leg. Hence, the signal controls for legs $W$ and $E$ are $S_{\{i,l,W,green\}}$ and $S_{\{i,l,E,green\}}$ in $l$ lanes, $S_{\{i,fr,W,red\}}$ and $S_{\{i,fr,E,red\}}$ in $fr$ lanes. All other phases in the $N$ and $S$ legs are in a red light situation: $S_{\{i,fr,N,red\}}$ and $S_{\{i,l,N,red\}}$ for the $N$ leg, $S_{\{i,fr,S,red\}}$ and $S_{\{i,l,S,red\}}$ for the $S$ leg.

Phases 3 and 4 are similar to Phases 1 and 2, but operate in the opposite direction.

## 6 Path Planning for Drones

The intersection network is continuously monitored by four drones following a predefined flight path. These drones operate within distinct rectangular routes, as shown in Fig. 6. The black arrows represent the drone movement directions within each route. Furthermore, the arrows indicate the parts of the road network included in the monitoring area, while other parts are outside the monitoring area. Black circles and black stars, highlight the flight points and meeting points assigned to each drone, respectively. Drones fly over the cells at different speeds, collecting data from four subcells in each direction. When each drone arrives at a meeting point, it transfers its most recent data related to its assigned path and neighbouring routes to the traffic agent. In this approach, information exchange happens in two directions: from the drone to the traffic agents and from the traffic agent to the drones. As a specific example, the last updated information relating to west lanes of agent number 5 collected by west drone is transferred to agents number 4, 1, and 2 by north drone. In addition, agent number 5 receives the north information from the drone and shares that with other drones in the west, east, and west directions. Therefore, the number of vehicles recorded by drones is utilized as the state space of the DQN in each agent to determine the most effective actions to minimize the overall cost of the city network.

### 6.1 Assumptions

In this simulation, some assumptions are considered to enhance the efficiency of the simulation as follows:

- Physical aspects of drones, such as battery life, weather condition, and communication costs, are not considered in this study.
- The cycle time of one intersection is determined by the sum of the green light durations for all traffic signals in a traffic unit as $C$.
- The yellow phase duration of all traffic signal remains fixed at $2s$.
- We define the minimum and maximum duration of the green traffic light as $S_{\{i,fr,k,green\}}min$, $S_{\{i,fr,k,green\}}max$ for $fr$ lanes and $S_{\{i,l,k,green\}}min$, $S_{\{i,l,k,green\}}max$ for $l$ lanes.

## 7 Simulation Result

In this section, we evaluate the designed strategy by studying congestion levels based on the remaining vehicle count on the connected roads at intersection 5 at the end of the assigned traffic phase, both with fixed control traffic light timing and with the adaptive control traffic light timing using DQN model. Since the DQN model has demonstrated superior performance in traffic management [3] compared to other strategies, we do not include comparisons with alternative methods in this study. In the simulation, the drone is modeled with a constant speed of 16 m/s, represented by $S_{Drone}$. Note that, the maximum capacity of each cell is 10 vehicles for the $R_{\{i,IN_{fr},k\}}$ lane and 5 vehicles for the $R_{\{i,IN_l,k\}}$ lane. Following the approach in [7], the simulation parameters are considered as $m = 32$, $\alpha = 0.0001$, and $\gamma = 0.99/MSE$. The input rates for the network are applied from alphabet points as shown in Fig. 4 with fixed rates of 1800 and 900 $Veh/h$ entering $fr$ and $l$ lanes, respectively. The simulation begins with 24 and 12 vehicles randomly placed in the subcells of the $fr$ and $l$ lanes, respectively, across all intersection legs. Figure 7a and 7b illustrate the total number of vehicles remaining in the lane $R_{\{5,IN_{fr},W/E\}}$ and $R_{\{5,IN_{fr},N/S\}}$ at the end of phase 1 and 3 of the traffic light at intersection 5 during the fixed control traffic light timing (see Fig. 7a and 7b, yellow lines), and during the adaptive control traffic light timing (see Fig. 7a and 7b, blue lines). Both traffic light timing scenarios are evaluated for 50 cycles with each cycle lasting 56 s. In the fixed control traffic light timing, the green traffic light duration is fixed at 12 s for all four traffic phases. As the simulation begins with ($S_{\{i,fr,W/E,green\}}$), the number of vehicles remaining in the west/east directions is lower than in the north/south directions under fixed control traffic light, with an average of approximately 30 to 50 vehicles per cycle, respectively. This indicates that, the fixed control traffic light is ineffective in preventing and resolving traffic congestion, as the input and output rates of the roads are equal. The simulation results for the adaptive control traffic light timing (see Fig. 7a and 7b, blue lines) show a clear reduction in traffic congestion across all roads, bringing it into the normal range approximately 10 and 7 vehicles in both $fr$ lanes. Figure 7c and 7d illustrate the total number of vehicles remaining in lane $R_{\{5,IN_l,W/E\}}$ and $R_{\{5,IN_l,N/S\}}$ at the end of phase 2 and 4 of the traffic light at intersection 5 during the fixed control traffic light timing (see Fig. 7c and 7d, yellow lines), and during the adaptive control traffic light timing (see Fig. 7c and 7d, blue lines).

We clearly notice that the north/south direction shows a higher number of vehicles, averaging 25 vehicles compared to 15 vehicles in the opposite directions. This is another clear indication that, the fixed control traffic light is ineffective in preventing and resolving traffic congestion, as the input and output rates of the roads are equal. The simulation results for the adaptive control traffic light timing (see Fig. 7c and 7d, blue lines) show a clear reduction in traffic congestion across all roads, bringing it into the normal range approximately 10 and 7 vehicles in both $l$ lanes. As a result, traffic congestion is influenced not only by the decisions of agent 5 but also by those of neighbouring agents, including agents 2, 4, 6, and 8. In general, traffic congestion can be controlled by adjusting

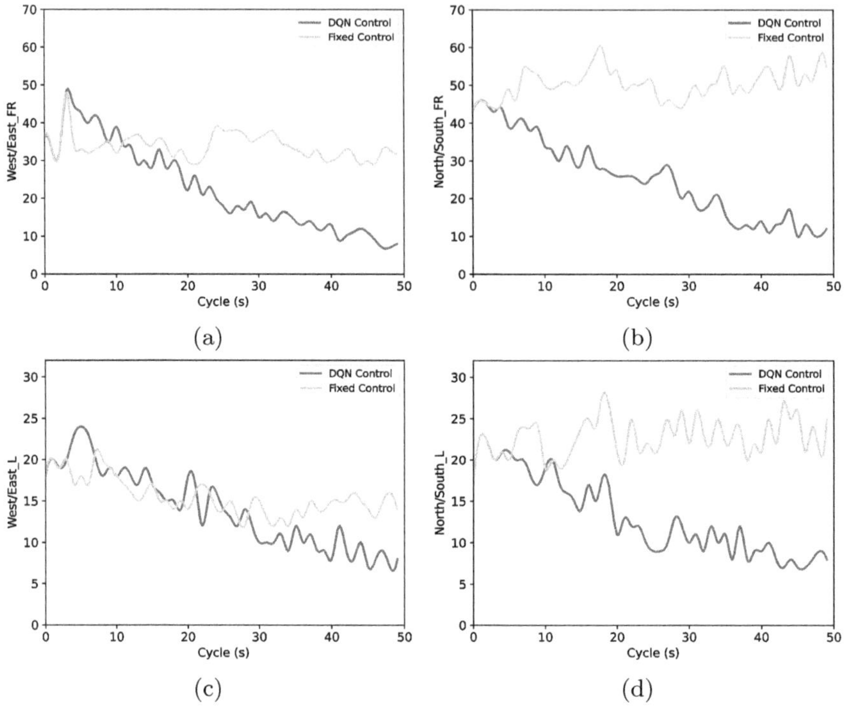


**Fig. 7.** The total number of vehicles remaining in the lanes of intersection 5 at the end of different green light phases. (Color figure online)

the rates of inflow and outflow. In our previous study [2], we controlled only the output rates while keeping the input rates constant, leading to traffic congestion fluctuations between the north/south and west/east directions. This indicates that the proposed method is not effective in minimizing the traffic congestion in both directions. In this study, we aim to monitor and manage traffic in both directions. For example, solving congestion in $R_{\{5,IN_{fr},W/E\}}$ depends on traffic phases 1 and 3 at intersections 2 and 8, as they impact the vehicle input rate on the west/east roads. Additionally, the road exit rate is regulated by extending the green time for phase 1 at intersection 5. To summarise, by leveraging a swarm of drones to transmit data from agent 5 to its neighbouring agents, the multi-agent learning strategy can optimize green traffic light durations at intersection 5 to minimize the remaining number of cars in interlinked roads. This adaptive approach adjusts for both input and exit traffic rates, providing a more adaptive solution than fixed control methods.

## 8 Conclusion

In this study, we designed multi agent deep reinforcement learning-based traffic control system by deploying a swarm of drones to tackle multi-intersection traffic optimization problems. Vissim simulator is used to model a real-world traffic grid-like scenario within nine intersections. Each intersection is trained separately using the generated data to reduce the number of vehicles remaining on the interlinked road at the middle intersection. The vehicle data is transferred from Vissim to a cellular map, where four drones monitored traffic and send updates to a central traffic unit at designated meeting points. In addition, the drones receive recent data from traffic units about other parts of the city regions under other drones surveillance and facilitated data exchange between intersections. The results showed that adaptive control traffic light timing based on DQN and drones surveillance for data collection significantly reduces congestion in all directions at the middle intersection compared to fixed control traffic light timing. This work highlights the potential of drone technology in improving autonomous traffic management in multi-intersection network. Future work will extend this approach to larger and more realistic urban networks.

## References


1. Al-Msari, H., Koting, S., Ahmed, A., El-Shafie, A.: Review of driving-behaviour simulation: vissim and artificial intelligence approach. Heliyon **10** (2024)
2. Alahvirdi, D., Tuci, E.: Intelligent traffic monitoring and management by swarm of drones and smart traffic units. In: Under review at IEEE Transactions on Intelligent Transportation Systems (2025)
3. Almukhalfi, H., Noor, A., Noor, T.H.: Traffic management approaches using machine learning and deep learning techniques: a survey. Eng. Appl. Artif. Intell. **133**, 108147 (2024)
4. Buzachis, A., Celesti, A., Galletta, A., Fazio, M., Fortino, G., Villari, M.: A multi-agent autonomous intersection management (MA-AIM) system for smart cities leveraging edge-of-things and blockchain. Neurocomputing **522**, 148–163 (2020)
5. Ge, H., Song, Y., Wu, C., Ren, J., Tan, G.: Cooperative deep Q-learning with q-value transfer for multi-intersection signal control. J. Control Sci. Eng. **7**, 40797–40809 (2019)
6. Gu, X., Abdel-Aty, M., Xiang, Q., Cai, Q., Yuan, J.: Utilizing UAV video data for in-depth analysis of drivers' crash risk at interchange merging areas. crash risk at interchange merging areas. Accid. Anal. Prev. **123**, 159–169 (2019)
7. Haddad, T.A., Hedjazi, D., Aouag, S.: A deep reinforcement learning-based cooperative approach for multi-intersection traffic signal control. Eng. Appl. Artif. Intell. **114**, 105019 (2022)
8. Li, X., Zhao, Z., Liu, L., Liu, Y., Li, P.: Multi-intersection traffic optimisation: an optimization model of multi-intersection signal control for trunk road under collaborative information. J. Control Sci. Eng. **1**, 2846987 (2017)
9. Mustapha, A., Abdul-Rani, A.M., Saad, N., Mustapha, M.: Advancements in traffic simulation for enhanced road safety: a review. Simul. Modell. Pract. Theory, 103017 (2024)
10. Nguye, D.D.: Cloud-based drone management system in smart cities. development and future of internet of drones (IoD). Insights, Trends Road Ahead **141**, 211–230 (2021)
11. Outay, F., Mengash, H.A., Adnan, M.: Applications of unmanned aerial vehicle (UAV) in road safety, traffic and highway infrastructure management: recent advances and challenges. Transp. Res. Part A: Policy Pract. **141**, 116–129 (2020)

12. Pham, H.Q., Camey, M., Pham, K.D., Pham, K.V., Rilett, L.R.: Review of unmanned aerial vehicles (UAVs) operation and data collection for driving behavior analysis. In: CIGOS 2019, Innovation for Sustainable Infrastructure. Crash Risk at Interchange Merging Areas. Accident Analysis & Prevention, pp. 1111–1116 (2020)
13. Roldán-Gómez, J.J., Garcia-Aunon, P., Mazariegos, P., Barrientos, A.: Warmcity project: monitoring traffic, pedestrians, climate, and pollution with an aerial robotic swarm: data collection and fusion in a smart city, and its representation using virtual reality. Pers. Ubiquit. Comput. **4**, 1151–1167 (2022)
14. Septyaningrum, I.M., Anindita, R.Y.: Traffic signalizing application at unsignalized intersection applying vissim software microsimulation. In: RSF Conference Series: Engineering and Technology, vol. 2, no. 2, pp. 294–306 (2022)
15. Wang, H., Chen, H., Wu, Q., Ma, C., Li, Y.: Multi-intersection traffic optimisation: a benchmark dataset and a strong baseline. IEEE Open J. Intell. Transp. Syst. **3**, 126–136 (2021)
16. Wang, Y., Liu, Y., Chen, W., Ma, Z.M., Liu, T.Y.: Target transfer Q-learning and its convergence analysis. Neurocomputing **392**, 11–22 (2020)
17. Wei, H., et al.: Learning max pressure control to coordinate traffic signals in arterial network. In: Proceedings of the 25th ACM SIGKDD International Conference on Knowledge Discovery and Data Mining, pp. 1290–1298 (2019)
18. Zaatouri, K., Ezzedine, T.: A self-adaptive traffic light control system based on YOLO. In: 2018 International Conference on Internet of Things, Embedded Systems and Communications (IINTEC), pp. 16–19 (2018)
19. Zhu, P., et al.: Detection and tracking meet drones challenge. IEEE Trans. Pattern Anal. Mach. Intell. **44**(11), 7380–7399 (2021)

# Accelerating Landslide Temporal Forecasting Through a Re-engineered Software System: $^{GA}$SAKe 2.1

Alessandro Sidero[1], Donato D'Ambrosio[1](✉), Valeria Lupiano[2], and Giulio Iovine[2]

[1] Department of Mathematics and Computer Science (DeMaCS), University of Calabria, 87036 Rende, Cosenza, Italy
donato.dambrosio@unical.it

[2] Institute of Research for Geo-Hydrological Protection (CNR-IRPI), National Research Council, 00185 Rende, Cosenza, Italy

**Abstract.** $^{GA}$SAKe is a linear, steady-state hydrological model employing Genetic Algorithms to predict the timing of landslide triggering based on rainfall and previous activations. Its calibration yields discrete kernels, enabling detailed analysis of even complex geohydrological conditions. Once validated, the model's mobility function allows for the prediction of landslide activations.

Whereas the 2.0 release introduced a renewed computational engine based on the ParadisEO version 2 software library, the latest 2.1 version of $^{GA}$SAKe, here presented, has undergone a complete re-engineering and is based on ParadisEO version 3. This comprehensive architectural revision delivers vastly superior computational performance compared to its predecessors and enables efficient parallel execution on shared-memory systems.

An application to the Uncino landslide in Southern Italy, characterized by six reactivations, is described to demonstrate the model's capabilities. Calibration of the re-engineered $^{GA}$SAKe confirmed the same optimal fitness results obtained with the previous version, now achieved with significantly enhanced efficiency, ranging from 4.04x to 8.0x in terms of speed-up, depending on the adopted fitness function out of the three available.

**Keywords:** Genetic Algorithms · ParadisEO 3.0 · Distributed Memory Execution · Hydrological Model · Prediction · Landslide Activation

## 1 Introduction

A vast portion of world's territory is affected by severe slope instability. According to [20], rainfall is responsible for about 90% of worldwide landslide-related casualties. About 500,000 slope movements were identified in Italy, with a mean of 1.6 failures per km$^2$ [13, 29]. Here, about 70% of landslides are due to rainfall [1]. In fact, temporary alterations in groundwater dynamics (due to water infiltration) influence slope instability [30], which often activate after the most intense phases of rainfall events.


M. Bianchini et al. (Eds.): WIVACE 2025, CCIS 3001, pp. 31–44, 2027.
https://doi.org/10.1007/978-3-032-33185-4_3

Two approaches are commonly employed to model the relations between rainfall and slope instability. The complete, or physically based approach, seeks to ascertain the impact of rainfall on slope stability by modelling its effects on overland flow, groundwater infiltration, pore pressure, and the associated balance of shear stress and resistance (cf. e.g., [18, 22, 28, 31]). The major drawback of this approach is the need of a substantial and expensive collection of detailed data for numerical modelling purpose. The hydrological approach [6] consists on a Machine Learning-like procedure that relies on the statistical-probabilistic analysis of rainfall series and the dates of slope-movement occurrences (see, among others, [3–5, 11–13, 23]).

$^{GA}$SAKe (Genetic-Algorithm based Self Adaptive Kernel) is a discrete hydrological model able to forecast the activation of slope movements [26]. The model, which is an evolution of $^{CM}$SAKe [25], is versatile and can be utilized for either individual landslides or a collection of comparable phenomena within a geographically uniform region. In a recent work [8], the release 2.0 of $^{GA}$SAKe was released, which represents a substantial enhancement of the original version [26]. This updated release was distinguished by: (i) multi-thread parallel execution capabilities, thanks to the adoption of ParadisEO 2.0 [15], (ii) a validation tool, designed to quantify the model's capacity to predict new activations based on several distinct criteria to assess the outcomes through different fitness functions, (iii), a non-linear regression tool that can approximate the model's discrete kernel by means of analytical expressions (such as linear functions and/or gamma distributions) to provide insights into groundwater dynamics leading to slope instability, and (iv) a reach graphical user interface with advanced visualization features.

In this paper, we introduce 2.1 version of $^{GA}$SAKe. The model has undergone complete re-engineering and is now based on ParadisEO version 3.0 [9]. This comprehensive architectural revision delivers vastly superior computational performance compared to its predecessors, enabling efficient parallel execution on shared-memory systems. An application to the Uncino landslide in Southern Italy – i.e., the benchmark application already adopted in [27] – characterized by six reactivations, is described to demonstrate the model's correctness (with respect to its predecessor) and improved computational performance.

The rest of the paper is structured as follows: Sect. 2 describes the $^{GA}$SAKe hydrological model, including modelling concepts, the genetic algorithm and the new software system architecture; Sect. 3 outlines the case study (benchmark); Sect. 4 presents computational results; Sect. 5 offers a discussion, by outlining possible future development and concluding remarks.

## 2 The $^{GA}$SAKe 2.1 Hydrological Model

$^{GA}$SAKe [8, 26] is a linear, steady-state hydrological model that uses Genetic Algorithms [14] to predict the timing of landslide activation based on rainfall data and past events. In the following, formalization of fundamental concepts, details of the genetic algorithm and of the re-engineered system software are briefly described.

## 2.1 Hydrological Modelling

In hydrology, the transformation of rainfall in runoff can be described by an integral of convolution between a unitary impulsive response of the basin – the kernel, $h(t)$ – and the rainfall, $p(t)$, assuming a linear, steady scheme [7]. The kernel represents the unitary volume influx in an infinitesimal period, and is defined as:

$$\int_0^\infty h(t)\mathrm{d}t = 1$$

in which $h(t) = h(-t)$, $h(t) \geq 0$.

In slope-stability analyses, the lower bound $t = 0$ corresponds to the landslide activation. The kernel assumes a finite duration, $t_b$, referred to as "base time", and refers to the "preceding rainfall" that is significant for the activation of a given landslide. In other words, rains older than $t_b$ can be neglected.

The integral of convolution is expressed as:

$$z(t) = \int_0^{t_b} h(\tau)p(t - \tau)\mathrm{d}\tau$$

in which $z(t)$ represents the "mobility function" at time $t$. For a specific case study, the kernel can be determined by means of calibration procedures in a Machine-Learning like manner, by relating landslide activations to rains.

In SAKe, a discretized approach is adopted: the elements of the kernel are characterized by width $\Delta t$ and height $h_i$, and the above equation can be re-written as:

$$z_u = \sum\nolimits_{i=1}^{n_{t_b}} h_i p_{u-i+1} \Delta t$$

where $n_{t_b} = t_b/\Delta t$ is the number of time intervals (of duration $\Delta t$) in which the base time is discretized, $\sum\nolimits_{i=1}^{n_{t_b}} h_i = 1$, and $u \in \mathrm{U} = \{u_1, u_2, \ldots, u_n\}$, $u \geq n_{t_b}$, , is an index corresponding to a discrete time interval (of duration $\Delta t$) in the range $\left[t_0, t_f\right]$, where the rainfall has been measured/forecasted. The function $f_m : U \to \mathbb{R}$ s.t. $f_m(u) = z_u$ is referred to as "discrete mobility function".

The rain series needed to analyse a given case study, characterized by a set of historical activations, must cover a period that includes the first and the last dates. More in detail, a set of at least $n_{t_b}$ rain values, recorded before the first date of activation, is needed to compute the mobility function.

## 2.2 The Genetic Algorithm

The genetic algorithm employed in $^{GA}$SAKe is characterized by a fixed population. Starting from an arbitrary set of $K$ kernels, a family of $Q$ optimal solutions can be iteratively obtained, and quantitatively evaluated by a fitness function. Based on such kernels, a set of mobility functions is obtained, allowing to predict the critical conditions leading to instability. The genetic algorithm, described below, offers several options concerning the initial population, the fitness function, the selection of operators, and the replacement of population.

**Input and Initial Shape.** As concerns the input, the rainfall series and the dates of landslide occurrence must be provided. Rainfall can be expressed at any time aggregation (e.g. weekly, daily, hourly), allowing landslides of very different sizes to be studied. Dates usually refer to known historical re-activations of a given slope movement. Alternatively, they may refer to events of shallow-landslide activations in a given, homogeneous study area.

Concerning the initial shape for the kernels, they can be selected among a set of geometrical prefixed types (rectangular, decreasing triangular, increasing triangular) or left random. A rectangular shape implies equal weights of the rains within the base time before activation, whereas a decreasing/increasing triangular shape allows to assign greater/smaller weights to recentmost rains.

The base times, $t_b$, are randomly initialized in a suitable range $[t_{b_{min}}, t_{b_{max}}]$. Commonly, $t_b$ ranging from a few hours to some weeks, are suggested for shallow landslides, while greater values (up to several months) are suitable for larger phenomena.

**Fitness Functions and Additional Criteria.** As concerns the fitness, $\varphi$, three distinct functions are available. Values are normalized to the theoretical achievable maximum value, for comparison purposes. Additional criteria can be employed to choose among kernels with similar fitness values.

*Hierarchically Weighted (HW).* The $N$ mobility function's local maxima are ranked in decreasing order as $r_i = i$, with $i = 1, 2, \ldots, N$. Peaks that do not correspond to any activation have a null contribution to the fitness. In the other cases, maxima contribute to the individual fitness by an amount $w_i = 1/r_i = 1/i$.

For instance, let us consider the $k^{th}$ mobility function, $f_{m_k}$. If $L = 2$ dates of landslide activation are available, and both correspond to the two highest peaks (ranks $r_1 = 1$ and $r_2 = 2$), the obtained fitness of the $k^{th}$ kernel is $\varphi_k = 1 + 1/2$. On the other hand, if for another mobility function $f_{m_l}$ the two activation dates are captured by the first and fifth ranked peaks (ranks $r_1 = 2$ and $r_5 = 5$), the fitness for the corresponding $l^{th}$ kernel is $\varphi_l = 1 + 1/5 < \varphi_k$.

*Equally Weighted (EW).* The $N$ mobility function local maxima are ranked in decreasing order as $r_i = i$, with $i = 1, 2, \ldots, N$. Peaks that do not correspond to any activation have a null contribution to the fitness. In the other cases, maxima contribute to the individual fitness by an amount $w_i = 1/L$ if $i \leq L$, and of $w_i = 1/i$ if $i > L$.

For instance, let us consider the $k^{th}$ mobility function, $f_{m_k}$. If two dates of landslide activation are available and both are found in correspondence of the two highest peaks (ranks $r_1 = 1$ and $r_2 = 2$), the obtained fitness for the corresponding $k^{th}$ kernel is $\varphi_k = 1/2 + 1/2 = 1$. On the other hand, if for another mobility function $f_{m_l}$ the two activation dates are captured by the first and fifth ranked peaks (ranks $r_1 = 1$ and $r_5 = 5$), the fitness for the corresponding $l^{th}$ kernel is $\varphi_l = 1/2 + 1/5 < \varphi_k$.

*Receiver Operating Characteristic (ROC) Curve.* In Machine Learning, a ROC curve is a graphical representation of the performance of a classification model. It plots the true positive rate against the false positive rate at various threshold settings. A prefixed number ($n_{tr}$) of thresholds, $z_1, z_2, \ldots, z_{n_{tr}}$, is assumed, each threshold $z_k$ being expressed with reference to the smallest peak (relative maximum $z_M$) of the mobility function:

$$z_k = k\frac{z_M}{n_{tr}}(1 < k < n_{tr})$$

By considering such thresholds, the peaks of the mobility function can be distinguished into 4 classes, based on their correspondence with the $L$ dates of activation. For each threshold, the following elements of the contingency table can be determined:

- *TP* (True Positive), peaks $z_j \geq z_k$ and coincide with one activation.
- *TN* (True Negative), peaks $z_j < z_k$ and do not coincide with any activation.
- *FP* (False Positive), peaks $z_j \geq z_k$ and do not coincide with any activation.
- *FN* (False Negative), peaks $z_j < z_k$ and coincide with one activation.

The above elements, as obtained for all the $n_{tr}$ thresholds, allow to define a set of points given by:

$$(1 - specificity, sensitivity) = \left(1 - \frac{TN}{FP + TN}, \frac{TP}{TP + FN}\right)$$

A ROC curve can be drawn, based on the above points. The ROC fitness of the kernel is evaluated by the Area Under the Curve (AUC): the worst performances are characterized by a value equal to 0.5, whereas the best ones have values 1.

*Secondary Fitness Criteria.* In case of kernels with similar fitness values, the best can be selected by one of the following additional criteria (or by a combination of them):

1. Minimizing the base time, $t_b$.
2. Maximizing the "safety margin", defined as $\Delta z_{cr} = (z_{min} - z_{cr})/z_{min}$, being $z_{min}$ the mobility function lowest peak corresponding to an activation date, and $z_{cr}$ the critical threshold, namely the highest peak just below $z_{min}$.
3. Minimizing the first-order momentum of the kernel with respect to the vertical axis: $\mu_0 = \sum_{i<t_b}(i - 0.5)h_i\Delta t$.

The above criteria allow for the activation of early warning procedures with the greatest advance, generally corresponding to impulsive responses to rainfall. See [8] for further details.

**Selection.** Four selection criteria are available:

- *Stochastic Tournament.* Two kernels are selected at random, and the kernel with highest fitness is selected, according to a probability given by the tournament rate parameter $T_r$, $0.55 \leq T_r \leq 1$.
- *Deterministic Tournament.* In a set of $T_s$ randomly selected kernels, the one with highest fitness is selected. $T_s$ is referred to as tournament rate.
- *Proportional (Roulette wheel).* A kernel is selected with a probability given by its fitness value over the sum of all individuals' fitness values.
- *Ranking.* The $K$ kernels are sorted in decreasing order, based on their fitness, and are assigned a rank$r_i = i, r_i \in \{1, 2, ..., K\}$. A kernel is selected by a roulette-like procedure, based on rank,$r$, population size,$K$, and selective pressure,$s$. Options available are:

- Linear ranking probability: $P_{lr} = \frac{1}{K}[s - 2(s-1)\frac{r-1}{K-1}], 1 \leq s \leq 2$
- Exponential ranking probability: $P_{er} = s^{r-1}\frac{1-s}{1-s^K}, 0 < s < 1$

**Genetic Operators.** The following operators are considered in $^{GA}$SAKe 2.1:

- *Crossover*. Depending on a prefixed crossover probability,$P_c$, two kernels having $t_{b_1}$ and $t_{b_2}$ base times are chosen among those selected (by the selection operator); a crossover point is randomly chosen in the range $]1, min(n_{t_{b_1}}, n_{t_{b_2}})[$ and kernels' pieces (before and after the crossover point) swapped to obtain two new offspring.
- *Mutation*. Depending on a prefixed mutation probability, $P_m$, the base time and the heights of a subset of kernel's elements are mutated. In particular:

  - The base time, $t_b$, is mutated, within the $[t_{b_{min}}, t_{b_{max}}]$ range, by multiplying the current value by a factor $d_{t_b} \in \left[1/P_{m_{t_b}}, P_{m_{t_b}}\right]$, where $P_{m_{t_b}} > 1$ is referred to as base time mutation parameter.
  - A random number of kernel's elements, $P_{me}$, expressed as % of $n_{t_b}$, is mutated, by multiplying their height, $h_i$, by a random factor $dh \in [P_{dH}\, h_{max}, P_{dH}/h_{max}]$, with $P_{dH}$ expressed as % of $h_{max}$.

Any new kernel is eventually normalized, to satisfy the continuity equation of volumes, $\sum_{i=1}^{n_{t_b}} h_i = 1$.

**Replacement.** As concerns replacement, the following types are allowed:

- *Elitist/steady state*. The best $n_e$ (elitist) kernels are selected from the current population and copied to the next iteration. The remaining $K - n_e$ kernels of the next population are then obtained by applying the above-mentioned Selection and Genetic Operators (note that, the $n_e$ kernels can also be selected to this purpose).
- *Generational*. A number of $K$ kernels are selected for reproduction, by applying the above-mentioned Selection and Genetic Operators, to generate the next population.

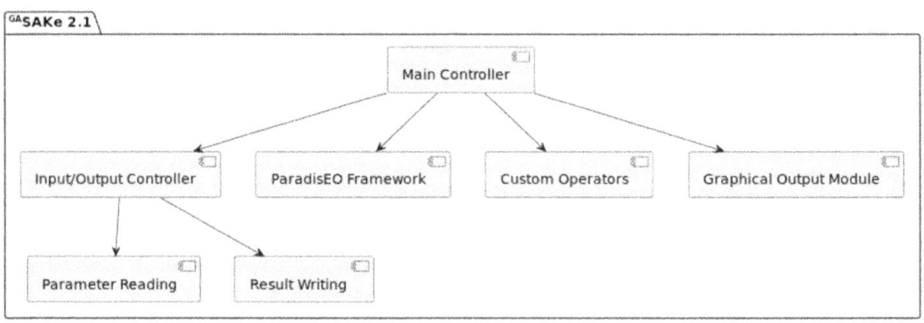


**Fig. 1.** System architecture of $^{GA}$SAKe version 2.1.

### 2.3 Re-engineering of $^{GA}$SAKe

$^{GA}$SAKe 2.0 introduced better computational performance and a rich graphical user interface with respect to its predecessor. Nevertheless, the strict correlation between computation and user interface could represented an obstacle to the further development of the system, besides slowing down the model execution due to the time wasted to update the user interface widgets. For this reason, a thorough re-engineering of $^{GA}$SAKe was performed in this paper, resulting in $^{GA}$SAKE 2.1. An elevated care was dedicated to the code quality, by adopting the principles of performance, modularity, independence among modules, and code maintainability.

The software architecture of SAKe 2.1 is shown in Fig. 1, which allows for a better components' separation with respect to $^{GA}$SAKe version 2.0. A main controller is responsible for the operations performed by the model. It controls the Input/Output module, the ParadisEO 3.0 framework, the custom operators (derived by the base ParadisEO's operators), and the graphical output module that is now based on Gnuplot and can be optionally activated during the model execution for interactive visualization purpose. Keeping the graphical module deactivated results in a command line execution for the model, permitting to do not waste time due to user interface's widgets update. This option was not available in GASAKe 2.0, due to its strictly coupled correlation between user interface and computing code.

One of the most impactful changes was the adoption of release 3.0 of ParadisEO for implementing the genetic algorithm. This new version introduces several improvements and optimizations, including preliminary support for execution on distributed memory systems. In addition to its usage for enhancing execution efficiency, ParadisEO 3.0 was also leveraged to standardize the input/output interfaces of the SAKe 2.1 model, contributing to a more modular and maintainable codebase.

Another component that underwent revision was the management of dates/time conversions, massively employed in each fitness function of the genetic algorithm. In version 2.0 of SAKe, this process relied on the `date_time` module of the Boost library [2], specifically using the `ptime_from_tm()` function to convert `tm` structures into `ptime` objects. This function accounted for approximately 35% of the fitness function's total execution time. To both reduce external dependencies and improve performance, the standard C++ library function `mktime()` was initially considered. This function converts a `tm` structure into a `time_t` value representing the number of seconds since January 1, 1970 (epoch time). However, due to internal validations, system structure accesses, and complex conversions, its repeated invocation introduced a significant performance overhead. Profiling results confirmed that `mktime()` was responsible for 88.97% of the fitness function's execution time, constituting a major bottleneck. To address this issue, the `sake_mktime()` custom function was developed. Designed to be deterministic, lightweight, and free of system calls, the new implementation reduced the function's contribution to the total execution time to 21.27%, significantly lowering the computational burden compared to its alternatives.

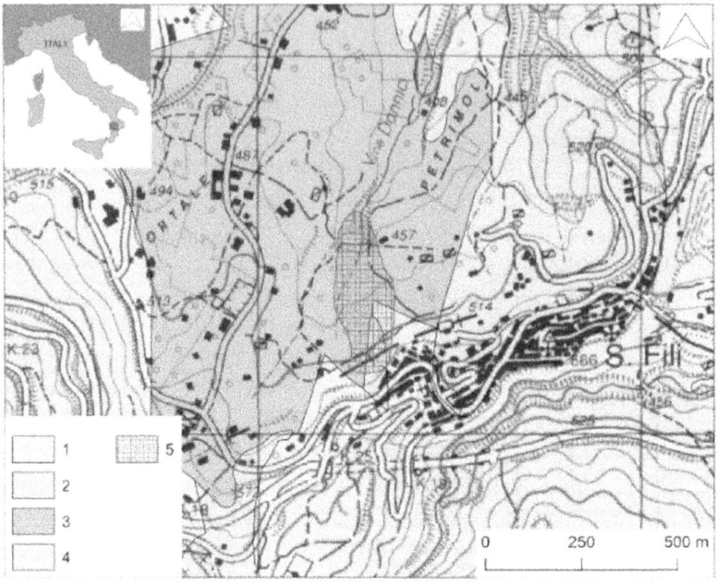


**Fig. 2.** Lithological sketch of the Uncino landslide. 1) Alluvium, colluvium, and residual soil; 2) conglomerate and sandstone; 3) Clay and clayey flysch; 4) Medium-high grade metamorphic rock. The landslide (5) is marked by a red hatched polygon. Lithological map after Geological map of Calabria - scale 1:25,000 (CASMEZ, 1967).

## 3 The Benchmark Study Case

This study applies $^{GA}$SAKe-2.1 to a case study in Northern Calabria, Italy, focusing on the Uncino landslide near San Fili. With an estimated volume of 3 million m$^3$, the landslide experienced six reactivations between 1960 and 1980. It lies at ~ 500 m a.s.l. on the NW margin of the village of San Fili, within the Crati graben, where weathered Palaeozoic metamorphic rocks are overlain by Miocene to Quaternary sediments (Fig. 2). The landslide mainly involved Miocene clay with minor sandstone above metamorphic bedrock. It is considered "intermediate" in deep-seatedness, measuring ca. 650 m in length, 200 m in width, and 25–30 m in depth.

Reactivations damaged local infrastructure, including roads and a railway. For modelling, only events from 1960–1980 were used; earlier ones lacked suitable data, and later events were excluded as the railway had been decommissioned. The closest rain gauge, in Montalto Uffugo (8 km north), is located in a region characterized by a wet season from October to March. Model calibration used daily rainfall from Sept 1, 1959, to Aug 31, 1980; validation covered Sept 1, 1980, to Mar 31, 1981.

**Table 1.** GA parameters adopted for the Uncino landslide experiments. Further details can be found in De Rango et al. (2025).

| Parameter Meaning | Value |
|---|---|
| Population size ($K$) | 100 |
| Maximum number of iterations | 10000 |
| Initial shape for the kernels | Decreasing Triangular |
| Selection type | Stochastic Tournament |
| Tournament rate ($Tr$) | 0.55 |
| Replacement type | Elitist/Steady State (10) |
| Crossover probability | 0.75 |
| Mutation probability ($P_m$) | 0.25 |
| Base time mutation parameter ($P_{m_{t_b}}$) | 50 |
| Minimum allowed value for the base time ($t_{b_{min}}$) | 15 |
| Minimum allowed value for the base time ($t_{b_{max}}$) | 90 |
| Number of kernel elements to be mutated ($P_{me}$), as $\%t_b$ | 25 |
| Kernel bars mutation parameter ($P_{dH}$), as $\%h_{max}$ | 50 |
| ROC lines | 10 |

**Table 2.** Convergence comparison in terms of GA iterations needed to reach the best fitness for each type of available fitness function.

| Fitness type | $^{GA}$SAKe 2.1 | $^{GA}$SAKe 2.0 |
|---|---|---|
| HW | 74 | 68 |
| EW | 50 | 59 |
| ROC | 83 | 115 |

# 4 Computational Results

All experiments were performed under WSL (Windows Subsystem for Linux) version 2.0 running Ubuntu 22.04 on a scientific workstation with the following specifications: AMD Ryzen 9 7950X 16-Core Processor (32-Thread), 32 GB of RAM, 1 TB SSD storage, and Windows 11 24.H2 operating system. The genetic algorithm configuration is shown in Table 1.

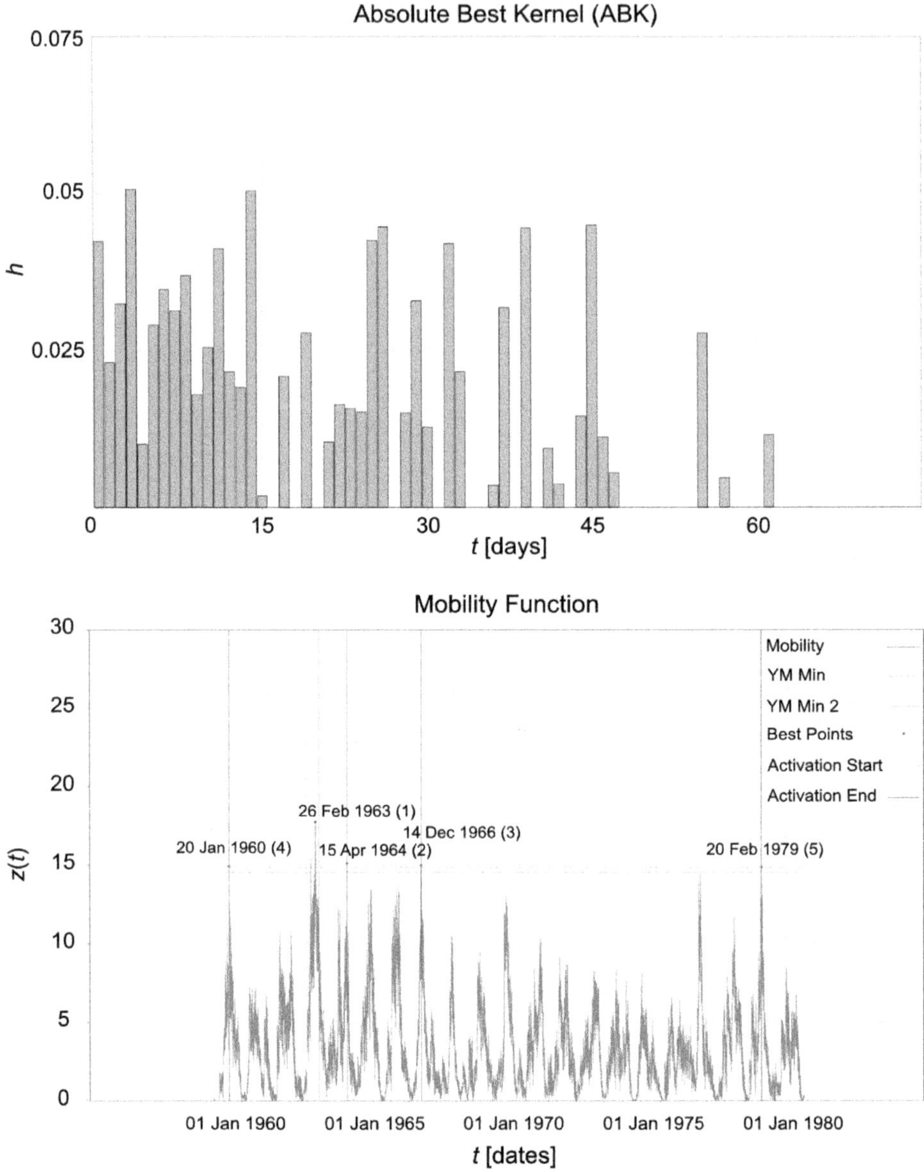


**Fig. 3.** Best kernel and corresponding mobility function for the Uncino study case.

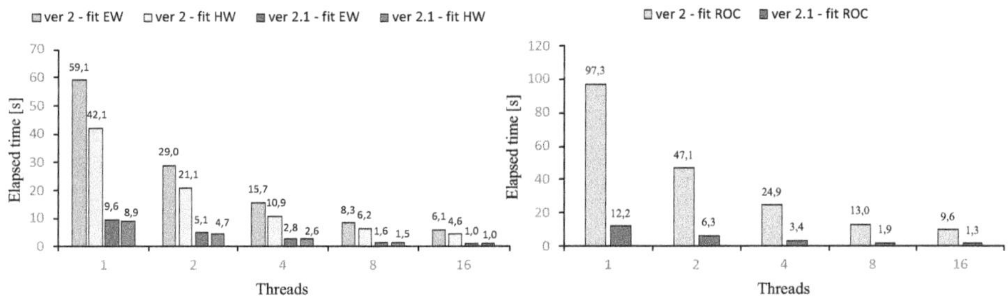


**Fig. 4.** Execution time comparison for increasing numbers of threads (1, 2, 4, 8, and 16) between $^{GA}$SAKe versions 2.0 and 2.1using the HW, EW, and ROC fitness functions. The elapsed times are reported on top of each bar.

The number of GA iterations needed to reach the best fitness is reported in Table 2. The observed differences are due to the new random generator engine of ParadisEO 3.0 (cf. [19] - ParadisEO 3.0.0—"Epimetheus", 2023 official changelog: "refactor `UF_random_generator` to fit new `std::shuffle`"). For both 2.0 and 2.1, the genetic algorithm rapidly converges to the best fitness in an order of magnitude of a hundred steps.

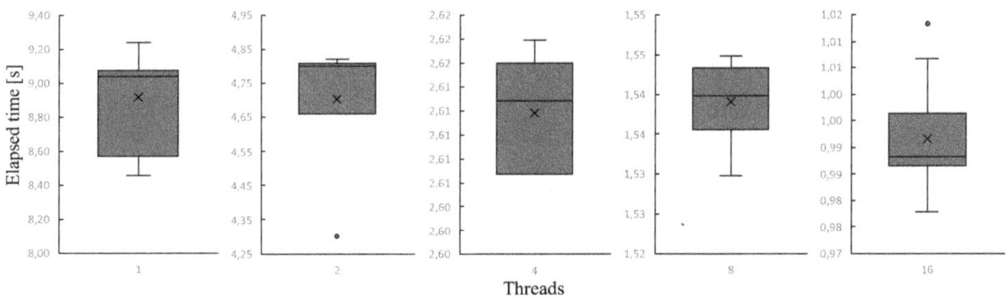


**Fig. 5.** Box plots showing the execution times over 10 runs of $^{GA}$SAKe versions 2.1 for increasing numbers of threads (1, 2, 4, 8, and 16) for the HW fitness function. Similar results were obtained for the EW, and ROC fitness functions.

For illustrative purpose, Fig. 3 displays the best kernel along with the corresponding mobility function of the best individual, obtained using the HW fitness function. As expected, the first five activations were correctly detected during calibration, with the sixth activation also accurately detected during the subsequent validation phase [8]. To assess the performance of $^{GA}$SAKe 2.1 with respect to $^{GA}$SAKe 2.0, a set of calibration experiments, each of them repeated ten times, were conducted by adopting the HW, EW, and ROC fitness functions, by varying the number of cores employed (namely 1, 2, 4, 8, and 16), and by disabling processor's Simultaneous Multi-Threading (SMT). Both versions were initialized by using the parameters reported in Table 1, and the same pseudo-random seed to ensure consistency of test conditions. Figure 4 summarizes the results and shows how version 2.1 was able to drastically reduce the total execution time in all the multithread scenarios. The speed-up registered ranged from about 4.04x, obtained with HW fitness function and 8 threads (the elapsed time was reduced from 6.2 to 1.54 s), to 8.0x, obtained with the ROC fitness function and 1 thread (the elapsed time was reduced from 97.3 to 12.17 s). Eventually, Fig. 5 reports the box plots showing the execution times over 10 runs of $^{GA}$SAKe versions 2.1 for increasing numbers of threads for the HW fitness function. Similar results, here omitted for the sake of conciseness, were obtained for the EW, and ROC fitness functions.

## 5 Discussion and Conclusion

In this paper, the 2.1 version of the $^{GA}$SAKe hydrological model is presented. Starting from the previous (2.0) release, the system software was completely rewritten, aiming at improving code quality and efficiency. The comparison between $^{GA}$SAKe 2.0 and $^{GA}$SAKe 2.1 clearly highlighted the operational advantages achieved through the re-engineering process. The adoption of modular architecture enabled a deep decoupling of the original system's components, thereby facilitating targeted optimization, maintenance, and testing. From a performance standpoint, the adoption of ParadisEO 3.0, the separation between computing and visualization code, and other optimizations introduced, drastically reduced execution times, ensuring a significant speed-up improvement (from 4.04x to 8.0x) in all the tested multi-thread scenarios.

The re-engineered codebase now provides a more suitable starting point for future model/software development. Some aspects could be improved or extended with respect to the current implementation. For instance, all the currently implemented fitness functions assign a null contribution to the local maxima of the mobilization function that do not correspond to any slope movement. This choice could be responsible for a propension for the genetic algorithm to select very complex kernels. Assigning negative contributions to peaks that do not match any activation date could result in more simple kernels, contextually improving the performance of the regression tool designed to go back to a continuous representation of the underground water dynamics [8]. Moreover, the present mutation operator may have a high impact on the kernel even when small values are adopted for its parameters. Finally, a variable population based evolutionary algorithm, like L-SHADE [24] and its more recent variations (e.g., [10, 16, 17, 21]), could be implemented, as such kind of models demonstrated to be able to provide better results with respect to fixed-population alternatives on a wide range of problems.

**Acknowledgments.** Donato D'Ambrosio is a member of the Gruppo Nazionale Calcolo Scientifico - Istituto Nazionale di Alta Matematica (GNCS-INdAM).

**Disclosure of Interests.** The authors have no competing interests to declare that are relevant to the content of this article.

## References

1. Alfieri, L., Salamon, P., Pappenberger, F., Wetterhall, F., Thielen, J.: Operational early warning systems for water-related hazards in Europe. Environ. Sci. Policy **21**, 35–49 (2012)
2. Boost Homepage. https://www.boost.org/. Accessed 10 June 2025

3. Brunetti, M.T., Peruccacci, S., Rossi, M., Luciani, S., Valigi, D., Guzzetti, F.: Rainfall thresholds for the possible occurrence of landslides in Italy. Nat. Hazard. **10**, 447–458 (2010)
4. Caine, N.: The rainfall intensity-duration control of shallow landslides and debris flows. Geogr. Ann. **62A**, 23–27 (1980)
5. Campbell, R.H.: Debris flow originating from soil slip during rainstorm in southern California. Q. J. Eng. Geol.Hydrogeol. **7**, 377–384 (1975)
6. Cascini, L. and Versace, P.: Eventi pluviometrici e movimenti franosi. In: Atti del XVI Convegno Nazionale di Geotecnica, Bologna, Italy, pp. 171–184 (1986). in Italian
7. Chow, V.T., Maidment, D.R., Mays, L.W.: Applied Hydrology, p. 572. Mc Graw Hill, New York (1988)
8. De Rango, A., et al.: $^{GA}$SAKe 2.0-an advanced hydrological model for predicting the timing of landslide activations. J. Environ. Inf. **45**(2), 84–102 (2025)
9. Dreo, J., et al.: from a modular framework for evolutionary computation to the automated design of metaheuristics—22 years of Paradiseo—. In: GECCO'21: Proceedings of the Genetic and Evolutionary Computation Conference Companion, pp. 1522–1530 (2021)
10. Fu, S., et al.: Modified LSHADE-SPACMA with new mutation strategy and external archive mechanism for numerical optimization and point cloud registration. Artif. Intell. Rev. **58**, 72 (2025)
11. Gariano, S.L., et al.: Calibration and validation of rainfall thresholds for shallow landslide forecasting in Sicily, Southern Italy. Geomorphology **228**, 653–665 (2015)
12. Guzzetti, F., Peruccacci, S., Rossi, M., Stark, C.P.: Rainfall thresholds for the initiation of landslides in central and southern Europe. Meteorol. Atmos. Physic **98**, 239–267 (2007)
13. Guzzetti, F., Peruccacci, S., Rossi, M., Stark, C.P.: The rainfall intensity-duration control of shallow landslides and debris flow: an update. Landslides **5**, 3–17 (2008)
14. Holland, J.H.: Adaptation in Natural and Artificial Systems. University of Michigan Press, (1975)
15. Keijzer, M., Merelo, J.J., Romero, G., Schoenauer, M.: Evolving objects: a general purpose evolutionary computation library. Artif. Evol. **2310**, 829–888 (2002)
16. Ma, J., et al.: A comprehensive comparison among metaheuristics (MHs) for geohazard modeling using machine learning: insights from a case study of landslide displacement prediction. Eng. Appl. Artif. Intell. **114**, 105150 (2022)
17. Mohamed, A.W., Hadi, A.A., Fattouh, A.M., Jambi, K.M.: LSHADE with semi-parameter adaptation hybrid with CMA-ES for solving CEC 2017 benchmark problems. In IEEE Congress on Evolutionary Computation (CEC), Donostia, San Sebastián, Spain, pp. 145–152 (2017)
18. Montgomery, D.R., Dietrich, W.E.: A physically-based model for the topographic control on shallow landsliding. Water Res. Res. **30**(4), 1153–1171 (1994)
19. Paradiseo 3.0.0—"Epimetheus" (2023). https://github.com/nojhan/paradiseo/releases/tag/v3.0.0#:~:text=refator%20UF_random_generator%20to%20fit%20the%20new%20std%3A%3Ashuffle
20. Petley, D.N.: The global occurrence of fatal landslides in 2007. In: International Conference on Management of Landslide Hazard in the Asia-Pacific Region, Tokyo, Japan, pp. 590–600 (2008)
21. Piotrowski, A.P.: L-SHADE optimization algorithms with population-wide inertia. Inf. Sci. **468**, 117–141 (2018)
22. Pisani, G., Castelli, M., Scavia, C.: Hydrogeological model and hydraulic behaviour of a large landslide in the Italian Western Alps. Nat. Hazard. **10**(11), 2391–2406 (2010)
23. Sirangelo, B., Versace, P.: A real time forecasting for landslides triggered by rainfall. Meccanica **31**, 1–13 (1996)

24. Tanabe, R., Fukunaga, A.S.: Improving the search performance of SHADE using linear population size reduction. In: IEEE Congress on Evolutionary Computation (CEC), pp. 1658–1665. IEEE, Beijing (2014)
25. Terranova, O.G., Iaquinta,25 P., Gariano, S.L., Greco, R. and Iovine, G.G.R.: $^{CM}$SAKe: a hydrological model to forecasting landslide activations. Landslide Sci. Pract. **3**, 73–79 (2013)
26. Terranova, O.G., Gariano, S.L., Iaquinta, P., Iovine, G.G.R.: $^{GA}$SAKe: forecasting landslide activations by a genetic-algorithms-based hydrological model. Geosci. Model Dev. **8**, 1955–1978 (2015)
27. Terranova, O., Gariano, S.L., Iaquinta, P., Lupiano, V., Rago, V., Iovine, G.: Examples of Application of $^{GA}$SAKe for predicting the occurrence of rainfall-induced landslides in Southern Italy. Geosci. **8**(2), 78–96 es8020078 (2018)
28. Terlien, M.T.J.: The determination of statistical and deterministic hydrological landslide-triggering thresholds. Environ. Geol. **35**, 125–130 (1998)
29. Trigila, A.: Rapporto sulle frane in Italia. Il Progetto IFFI. Metodologia, risultati e rapporti regionali, APAT, Roma, Italy, p. 681 (2007). in Italian
30. Van Asch, Th.V.J., Buma, J., Van Beek, L.P.H.: A view on some hydrological triggering systems in landslides. Geomorphology **30**(1–2), 25–32 (1999)
31. Wilson, R.C., Wieczorek, G.F.: Rainfall thresholds for the initiation of debris flow at La Honda, California. Environ. Eng. Geosci. **1**, 11–27 (1995)

# AILife: Artificial Intelligence in Life Sciences

# A Comparison on Cross–Domain Representation Learning for Skin–Lesion Classification

Paolo Andreini(✉), Simone Bonechi, Monica Bianchini, Franco Scarselli, Francesco Manni, Matilde Quercioli, and Barbara Toniella Corradini

Department of Information Engineering and Mathematics, University of Siena, Via Roma 56, 53100 Siena, Italy
paolo.andreini@unisi.it

**Abstract.** The advent of large–scale pre–training has produced powerful computer vision models, yet their effectiveness as feature extractors on highly specialized, out–of–distribution domains remains a critical issue. To fill this gap, we systematically compare the feature extraction capabilities of different, state–of–the–art architectures on the challenging task of dermoscopic skin–lesion classification—a domain substantially different from the natural images used during pre–training. We evaluate four distinct backbone architectures for feature extraction, encompassing both unimodal and multimodal models. Our selection includes EfficientNet B0 and Vision Transformer as representative unimodal models, along with Stable Diffusion Model and CLIP as state–of–the–art multimodal models. Using a subset of the ISIC archive, we assess the quality of extracted features via both linear probing and classification with a Multi Layer Perceptron. Our experiments suggest that all architectures produce highly effective features, achieving competitive performance. Notably, multimodal models, although pre–trained for tasks other than image classification, extract features that are remarkably competitive in this context. This work can serve as a comparative guide for the selection of feature extractors when tackling classification in specialized domains, such as medical imaging. The results highlight not only the generalizability of modern architectures, but also the surprising versatility of multimodal models as powerful feature extractors for interdisciplinary tasks.

**Keywords:** Skin lesion classification · Pre–trained backbones · Cross–domain representation learning · Feature comparison

## 1 Introduction

Melanoma is a global public health concern due to its aggressiveness and high mortality rate in case of late diagnosis (5–year survival >95% if caught early). Based on data collected by the International Agency for Research on Cancer[1],

[1] https://www.iarc.who.int.


M. Bianchini et al. (Eds.): WIVACE 2025, CCIS 3001, pp. 47–60, 2027.
https://doi.org/10.1007/978-3-032-33185-4_4

its distribution varies significantly by geographic location, ethnicity, and UV exposure levels.

Early–stage melanoma is highly treatable, but once it metastasizes, the prognosis worsens significantly. Misidentifying a melanoma as a benign nevus can delay necessary treatment and lead to worse outcomes. Indeed, some nevi may degenerate and show clinical and histological features similar to melanomas—such as asymmetry, irregular borders, and varied pigmentation. This resemblance increases the risk of misdiagnosis, either underestimating a melanoma or over–treating a benign lesion. Therefore, distinguishing between nevi and melanomas is critically important because it directly impacts early detection, accurate diagnosis, and effective treatment.

Traditional methods for diagnosing melanoma include visual examination of the skin, dermatoscopy (or epiluminescence), and in some cases, biopsy with histological examination. In the visual analysis, the dermatologist evaluates the suspicious skin lesion, observing the ABCDE characteristics of melanoma: Asymmetry, irregular Borders, non–uniform Color, Diameter greater than 6 mm, and Evolution over time. Dermatoscopy is a non–invasive test that uses a polarized light microscope to examine the skin, allowing the detection of abnormalities and irregularities in the pigmentation of the suspicious lesions. Finally, if dermatoscopy suggests possible melanoma, a biopsy is performed, taking a sample of the lesion for microscopic examination, which allows the presence of tumor cells to be confirmed or excluded. Moreover, diagnostic imaging (CT, PET–CT, ultrasound) may be necessary in some cases to evaluate the stage of melanoma and the spread of any metastases.

However, the subjective nature of visual assessment and the subtle differences between benign and malignant lesions can lead to diagnostic uncertainty, especially in early–stage melanomas or atypical nevi. To address these challenges, deep learning has emerged as a powerful tool for image–based medical diagnosis. By learning complex patterns and representations directly from dermoscopic or histopathological images, deep learning models can achieve diagnostic performance comparable to, and in some cases exceeding, that of experienced dermatologists.

Traditionally, models in medical image applications were trained from scratch or fine–tuned directly on the target domain [2,11,30,34], relying heavily on labeled data and domain–specific supervision. More recently, however, the success of machine learning and deep learning has shifted attention toward large–scale pre–trained models that demonstrate strong generalization across a wide range of visual tasks.

These models, originally trained on huge natural image datasets [35] with minimal supervision, are capable of learning dense and informative latent representations that transfer effectively to new domains—e.g., medical imaging. Among them, multimodal architectures such as Contrastive Language–Image Pretraining (CLIP) [31] and Stable Diffusion Model (SDM) [33] can learn even richer representations by jointly modeling visual and textual modalities. These models have shown promising results in downstream medical tasks such as classification [47], retrieval [26], and segmentation [17], without requiring fine–grained annotation or domain–specific supervision. Building on previous studies, this paper aims to answer the following question:

*Can multimodal models pre–trained on natural images help in skin lesion classification?*

To answer this question, we exploit latent representations produced by state–of–the–art multimodal models, such as SDM and CLIP, along with those generated by unimodal architectures such as Vision Transformer (ViT) and EfficientNet B0. While earlier work has examined backbone effectiveness on natural image tasks [16], our analysis is focused on a specific unseen domain.

We assess the quality of the latent representations in two ways. On one hand, on top of the feature extractor, we apply a standard linear–probing protocol, training a linear classifier to distinguish benign skin lesions from melanomas. On the other hand, we place a small Multi Layer Perceptron (MLP) to test the nonlinearity of the decision boundary. In both settings, the downstream task of skin–lesion classification achieve strong accuracy, indicating that representations learned from large, general–purpose data can be readily adapted to dermatological tasks.

The rest of the paper is organized as follows. In the following section, we report recent literature on deep learning methods used to distinguish images of nevi and melanomas. Such methods may include, where available, additional information on clinical data. Section 3 describes the data and models used to address the classification problem, while Sect. 4 collects and discusses the experimental results. Finally, Sect. 5 gathers conclusions and future perspectives.

## 2 Related Work

Traditional approaches to skin lesion classification typically involve adapting pre–existing architectures through fine–tuning or complete retraining (see Sect. 2.1). The recent emergence of large multimodal models, however, has opened new avenues for leveraging their powerful latent representations directly in few–shot settings, i.e., minimizing the need for retraining (see Sect. 2.2). This shift is motivated by necessity: training such models demands vast datasets, making subsequent adaptation via fine–tuning or retraining computationally infeasible, particularly within the data–limited context of medical applications. In this study, in order to further reduce the computational cost for retraining, we evaluate the use of pre–trained multimodal models as simple feature extractors.

### 2.1 Unimodal Training–Based Approaches

Convolutional Neural Networks (CNNs) have revolutionized computer vision—dominating the field for over a decade—and their impact extends far beyond natural–image tasks. Leveraging transfer learning, CNNs are now routinely applied across diverse diagnostic domains, from medical image analysis to skin lesion detection. One notable approach is SCDNet [29], a VGG16–based deep learning framework designed for multiclass classification of dermoscopic images into melanoma, melanocytic nevi, basal cell carcinoma, and benign keratosis. Achieving an accuracy of 96.91%, this model highlights the potential of CNN architectures in tackling complex classification tasks across multiple lesion types. The utility of transfer learning has been extensively explored in [14]. By exploiting both VGG16 and VGG19 architectures, a classification accuracy of 98.18% has been reported for skin lesion detection, emphasizing the benefits of pre–trained models on large datasets. Similarly, the BigTransfer (BiT) framework [38], based on a ResNet backbone, demonstrated superior performance in classifying melanocytic nevi, further validating the role of large–scale image pretraining in enhancing downstream dermatologic tasks. Direct comparisons with clinical expertise have also reinforced the value of deep learning. In [7], using 4,204 biopsy–confirmed cases, deep neural networks outperformed board–certified dermatologists in melanoma classification, highlighting the diagnostic power and potential of AI–assisted tools in routine practice. Focusing specifically on nevi, the research reported in [43] was aimed at classifying pigmented nevi via deep learning techniques, achieving a sensitivity of 88.65% and specificity of 90.26%. This work not only demonstrates high diagnostic accuracy but also correlates well with histopathological findings, bridging the gap between image–based prediction and clinical confirmation. Lastly, a more recent approach [24], integrating adaptive fine–tuning and ensemble CNNs, has been proposed for automatic diagnosis of melanoma and non–melanoma skin cancers. By using a combination of multiple dermoscopy datasets, this model achieved state–of–the–art performance in both binary and multiclass classification tasks, indicating the importance of model customization for specific dermatologic conditions. Recent advancements have further expanded the capability of deep learning models through the integration of multi–modal data fusion (combining images with clinical data such as patient age, lesion location, and lesion history) [3–6].

Vision Transformers (ViTs) [12] have also emerged as powerful architectures in medical image analysis, offering strong performance across tasks such as classification, segmentation, and detection. Compared to CNNs, ViTs benefit from global self–attention mechanisms, allowing them to model long–range dependencies and subtle features—especially useful in high–resolution medical imaging. In [15], ViTs have been employed for skin lesion classification, reporting an accuracy of 92.79% and a melanoma recall of 56.10%. While slightly lower in recall for melanoma, ViT offers advantages in model interpretability and long–range feature representation, marking a shift toward non–CNN–based models in medical image analysis. A recent survey [36] demonstrates the applicability of ViTs across various modalities, including radiology, histopathology, and dermatology. In skin lesion analysis, several studies have applied ViTs to standard benchmarks. Himel et al. [18] proposed a ViT–based pipeline combined with Segment Anything [25] for lesion localization, achieving 96.15% classification accuracy on HAM10000. Lightweight variants such as SkinDistilViT [21] reduce model size significantly while preserving classification accuracy, enabling faster inference and deployment on edge devices. Hybrid models, such as Swin Transformer backbones and CNN–ViT hybrids, further improve lesion classification by combining local inductive biases with global attention mechanisms [27]. PanDerm [28], a medical foundation model, integrates multiple ViT variants and contrastive objectives to achieve state–of–the–art performance across diverse dermatological tasks.

## 2.2 Multimodal Few–Shot Approaches

In recent years, large–scale models such as Vision Transformers, CLIP, and diffusion architectures have surged to prominence, fueling breakthroughs across numerous fields. Their capacity to encode rich, transferable visual representations underlies an exceptional ability to generalize across domains. Beyond their original purpose in image synthesis, Diffusion Models have recently emerged as powerful tools for downstream tasks such as classification, segmentation, and robust inference under distribution shift. Derm–T2IM [23] fine–tunes Stable Diffusion via few–shot learning to produce synthetic dermatoscopic images, boosting CNN and ViT classifier accuracy when used alongside real data. Gated-SegDiff [41] integrates gated feature fusion and neighborhood attention during denoising to improve lesion boundary delineation, while DiffMIC [44] applies dual–guidance diffusion encoders and surpasses standard CNN and transformer baselines on HAM10000. CKDNet [22] diffuses knowledge between segmentation and classification subnetworks, achieving state–of–the–art results on ISIC'17/18 without external data. Prompt–guided diffusion for visual–text data augmentation further enhances segmentation metrics [13], and synthetic lesion images from diffusion models prove beneficial in low–data regimes [37]. Together, these works underscore the flexibility of multimodal pre–trained large models in addressing data scarcity, segmentation, domain shift, and interpretability.

Multimodal Contrastive vision–language models have emerged due to their ability to align visual and textual modalities in a shared semantic space. The CLIP framework [32], which pairs a ViT–based image encoder with a text encoder trained using a contrastive loss on image–text pairs, enables zero– and few–shot classification by aligning lesion images with class–descriptive prompts—reducing the need for labeled data. Its adaptation to medical imaging, including dermatology, yields robust sample–efficient performance on HAM10000 and ISIC datasets [49]. MediCLIP demonstrates improved anomaly localization with limited examples [46], while hierarchical prompt–based variants achieve strong results in chest X–ray and skin–lesion classification tasks [10, 45]. MedCLIP [48] and BiomedCLIP [42] fine–tune CLIP on domain–specific datasets (e.g., MIMIC–CXR, PubMed, HAM10000), demonstrating improved performance on classification and retrieval tasks with limited supervision. In dermatology, prompt–engineered CLIP models have achieved competitive accuracy on skin lesion classification and retrieval tasks without fine–tuning, highlighting their generalization potential.

## 3 Materials and Methods

As detailed in Sect. 3.1, in this study we make use of two unimodal models, commonly used as feature extractors, and two multimodal models, known for their rich latent representations learned from large-scale pretraining. These models were chosen primarily for their strong and well-documented encoding capabilities, which have made them reference architectures in a wide range of visual and multimodal tasks, as well as for the availability of stable and well-maintained implementations. The skin–lesion dataset employed in our study is detailed in Sect. 3.2. Finally, the experimental setup used to evaluate the features extracted by various backbones is presented in Sect. 3.3.

### 3.1 Models

**Unimodal Models** *ViTs* [12] apply the transformer architecture, originally designed for natural language processing, to image data. Instead of operating on convolutional feature maps, ViTs divide an image into a sequence of fixed–size patches (e.g., $16 \times 16$ pixels). These patches are linearly embedded, enriched with positional encodings, and treated as input tokens to a standard transformer encoder. Each token attends to all others via multi–head self–attention, allowing the model to capture both local and global relationships within a unified representation. The global receptive field of self–attention makes ViTs especially suited for tasks where contextual reasoning is essential, such as identifying complex structures or subtle features in medical images. ViTs benefit significantly from large–scale pretraining, and their performance may degrade in low–data regimes unless adapted with hybrid approaches (e.g., Swin Transformers [27]) or regularization strategies. In this paper, we employ the Hugging Face open–source ViT implementation[2], which contains 86.6M parameters. We use the class token emitted by the last transformer encoder block, which serves as an image representation.

*EfficientNet* is a family of convolutional neural networks introduced in [40] that achieves state–of–the–art accuracy while reducing computational cost. Its key idea—called compound scaling—jointly scales network depth, width, and input resolution using a simple set of coefficients, rather than tuning each dimension independently. Starting from a highly optimized baseline model (EfficientNet B0), this balanced scaling strategy produces a series of models (B1–B7) that maintain an excellent accuracy–to–parameter ratio across a wide range of resource budgets. Because EfficientNet delivers comparable or better performance with far fewer parameters and FLOPs than predecessors like ResNet or Inception, it has become a widely used backbone for vision tasks. In this work, we use EfficientNet B0, which comprises 5.3M parameters, as a feature extractor, using the last convolutional feature map immediately before the classification head and flattening it into a single vector.

[2] https://huggingface.co/google/vit-base-patch16-224.

**Multimodal Models** *CLIP* [32] is a multimodal model that jointly learns visual and textual representations by aligning them in a shared embedding space. It is composed of two separate encoders: a vision encoder (often a ViT or ResNet) and a text encoder (typically a transformer–based language model). During training, CLIP receives batches of paired image and text descriptions and optimizes a contrastive loss that brings the embeddings of matching pairs closer while pushing apart non–matching ones. This results in a model that can directly compare and align an image with arbitrary text queries at inference time. In medical imaging, CLIP–style models have been adapted using domain–specific text corpora and visual datasets, enabling effective generalization even in low–data or cross–domain settings. In this paper, we leverage the Hugging Face open–source CLIP implementation[3], which comprises 151M parameters. The embeddings produced by the vision encoder, obtained after the final normalization layer that projects them into the shared multimodal space, are used as features.
*SDMs* [33] are generative models that synthesize data by reversing a gradual noising process. They extend Diffusion Models (DMs) [20] by introducing conditioning mechanisms [19], enabling the injection of textual or visual prompts via cross–attention. DMs consist of a *forward process*, which adds Gaussian noise to the input over multiple time steps, and a *reverse process*, which learns to denoise and recover the original data. The forward corruption follows a fixed noise schedule, gradually transforming clean inputs into near–isotropic Gaussian noise. As the reverse process is intractable, it is learned by a neural network—typically a U–Net—that predicts the noise at each step and models the data distribution. The model is trained with a simplified denoising objective to reconstruct the added noise and enable generation. Despite generating high–quality images, early models like DDPMs [20] operate directly in pixel space, resulting in prohibitive computational cost. DDIMs [39] mitigated this limitation by removing the Markovian assumption and adopting deterministic sampling, enabling faster generation with minimal loss in quality. Latent Diffusion Models (LDMs) [33] improve generative efficiency by shifting the diffusion process from high–dimensional pixel space to a compressed latent space obtained through a pre–trained autoencoder. This latent–space formulation not only reduces computational cost but also facilitates conditioning, allowing the integration of additional modalities such as text or images into the generative process. Originally intended for generative applications, SDMs have been found to capture semantically rich features through large–scale image–text training, achieving competitive results in downstream tasks like segmentation [1,41] and classification [44]. In this work we employ a pre–trained open–source implementation[4] of Stable Diffusion Model (v1.5). Features were extracted from the U–Net backbone, comprising 860M parameters, at $8 \times 8$ resolution and flattened into a one–dimensional vector.

[3] https://huggingface.co/openai/clip-vit-base-patch32.
[4] https://github.com/hkproj/pytorch-stable-diffusion.

### 3.2 ISIC Dataset

The International Skin Imaging Collaboration (ISIC)[5] is a joint academia and industry project designed to facilitate the application of digital skin imaging to help reduce melanoma mortality. In particular, the ISIC dataset is an open source repository of skin images collected to facilitate and promote researchers to develop automated diagnostic systems in this field. As of May 2025, the ISIC Archive—the largest publicly available collection of quality–controlled dermoscopic images—houses over 1.18 million images, approximately 521,000 of which are publicly accessible, with around 18,800 annotated for tasks like classification and segmentation. These images were sourced from multiple dermatologic centers worldwide and are accompanied by metadata such as age, sex, anatomical site, and clinical size. The dataset was designed to simulate lower–quality, non–dermoscopic images, making it particularly useful for training algorithms intended for primary care or non–clinical settings. In this study, we utilized images from the ISIC challenges held in 2017, 2019, and 2020. Duplicate images were removed following the methodology described in [9]. Since the 2017 and 2019 challenges were designed for multiclass classification tasks, we converted their labels to binary classification (melanoma vs. non–melanoma) to match the format of the 2020 challenge. The training set consists of 10,000 randomly sampled training images, and 2,000 images are used for validation (balanced between melanoma and normal classes). The original ISIC 2020 test set, comprising 10,982 images without publicly available annotations, is used. Consequently, all reported results were obtained through the official submission server for the ISIC 2020 challenge on Kaggle[6]. In ISIC 2020, about 30% of the hidden images were designated as the "public" test set for the live leaderboard, whereas the remaining ~70% comprised the "private" test set that determined the final rankings.

### 3.3 Experimental Setup

In this section, we describe the experimental setup and present the evaluation results. The goal is to compare the effectiveness of different backbone architectures for feature extraction from dermoscopic skin lesion images (see Fig. 1). We use frozen image encoders to extract features, comparing SDM, CLIP, ViT, and EfficientNet B0.

We evaluate the extracted features in two setups: linear probing and MLP classification. Linear probing evaluates the linear separability of the features by training a linear classifier on top of the frozen backbone representations, following established protocols in self–supervised learning [8,31]. This approach provides a direct measure of the quality of the learned representations and their alignment with the downstream classification objective. Instead, to evaluate non–linear decision boundaries, an MLP classifier, whose hyperparameters are determined using a grid search approach, is employed. The hyperparameter optimization is carried out by varying the number of hidden units

[5] https://www.isic-archive.com/.

[6] https://www.kaggle.com/c/siim-isic-melanoma-classification/.

**Fig. 1. Overview of our classification pipeline.** A dermoscopic image is encoded into the latent space of a frozen encoder. We extract the latent representation and give it as input to a classification module to distinguish between nevi and melanomas.

$h = \{128, 256, 512, 1024\}$, the dropout probabilities $d = \{0.0, 0.3, 0.5\}$, and the learning rate $\eta = \{10^{-3}, 10^{-4}, 10^{-5}\}$.

## 4 Results

In this section, we detail the experimental results of our comparative study. Table 1 reports the accuracy on the validation set obtained by employing the features extracted with different backbones using linear probing and the MLP–based classifier.

**Table 1.** Results on the validation set: (a) linear probing and (b) MLP–based classifier.

(a) Linear probing

| Model | Accuracy | Precision | Recall | ROC |
|---|---|---|---|---|
| SDM | 0,840 | 0,938 | 0,729 | 0,946 |
| EfficientNet B0 | 0,852 | 0,944 | 0,749 | 0,949 |
| ViT | 0,859 | 0,944 | 0,763 | 0,949 |
| CLIP | **0,871** | **0,948** | **0,785** | **0,955** |

(b) MLP–based

| Model | Accuracy | Precision | Recall | ROC |
|---|---|---|---|---|
| SDM | 0,873 | 0,954 | 0,783 | **0,961** |
| EfficientNet B0 | 0,863 | **0,956** | 0,762 | 0,954 |
| ViT | 0,869 | 0,920 | 0,808 | 0,947 |
| CLIP | **0,877** | 0,923 | **0,822** | 0,952 |

Across both setups, the CLIP backbone consistently produces the most discriminative features—leading in linear separability and top accuracy—while unimodal backbones (EfficientNet B0 and ViT) trail behind. Adding a small MLP head greatly boosts all models (especially SDM and ViT, which see the largest recall gains), yet MLP classification performance using CLIP features remain the highest overall, underlining the synergy between contrastive pre–training and lightweight non–linear classification. Interestingly, SDM—despite yielding the lowest scores under linear probing—experiences the biggest lift from the MLP head, jumping to an accuracy of 0.873 and achieving the highest ROC (0.962) of all MLP–based models. This suggests that SDM distilled features, while not immediately linearly separable, encode rich non–linear patterns that an MLP can effectively discern.

Table 2 reports the performance of the linear probing and the MLP–based classifier on the test set.

**Table 2.** Comparison of Feature Extractor Performance (ROC) on the ISIC Test Set Using Two Classifiers: (a) a Linear Probe and (b) an Optimized MLP.

(a) Linear Probing

| Model | Private | Public |
|---|---|---|
| SDM | 0.824 | 0.821 |
| EfficientNet B0 | **0.866** | 0.874 |
| ViT | 0.849 | 0.877 |
| CLIP | 0.855 | **0.892** |

(b)Best MLP results

| Model | Private | Public |
|---|---|---|
| SDM | 0.835 | 0.840 |
| EfficientNet B0 | 0.864 | 0.883 |
| ViT | 0.866 | **0.905** |
| CLIP | **0.877** | 0.897 |

These results confirm that CLIP consistently delivers the most discriminative features under linear probing and retains its lead when paired with an MLP head, underscoring the power of contrastive pre–training. While EfficientNet B0 and ViT also gain from non–linear adaptation—ViT, in particular, showing a large recall boost —, SDM experiences the largest relative lift from the MLP still remaining the weakest overall, highlighting that its distilled embeddings encode rich non–linear patterns but are intrinsically less linearly separable than those from larger CNN or CLIP.

## 5 Conclusions and Future Work

In this study, we assess the effectiveness of diverse models, trained on natural images, as frozen feature extractors for dermoscopic skin–lesion classification. By employing an SDM, the CLIP image encoder, a ViT, and EfficientNet B0, we show that representations learned on natural–image corpora transfer remarkably well to the specialized task of melanoma diagnosis. Across the ISIC benchmark, features from each backbone are evaluated by linear probing and shallow MLP classifiers, showing high ROC–AUC scores. Diffusion–derived features—despite being obtained from a generative model—are unexpectedly competitive with purely discriminative unimodal models. These findings underscore that modern architectures learn semantically rich latent spaces whose geometry remains robust under substantial domain shifts. Future work should expand the suite of feature extraction methods and integrate finer–grained lesion taxonomies and clinical metadata, moving beyond binary melanoma versus nevus classification toward realistic multi–label scenarios that encompass rare subtypes. This work lays the groundwork for employing pre–trained backbones in highly specialized medical image domains. Moreover, expanding our evaluation to multiclass lesion taxonomy and pixel–level tasks (e.g., segmentation) will test representation generality across clinical workflows. Incorporating non image metadata (e.g., patient demographics, lesion history) into multimodal diagnostic models could enhance accuracy and interpretability. These avenues aim to advance reliable, scalable AI solutions in dermatology.

## References


1. Amit, R., Mehta, V., Cao, L., Wang, S., Liang, P., Fei-Fei, L.: SegDiff: segmenting medical images with denoising diffusion models. In: Proceedings of the NeurIPS Workshop on Medical Imaging Meets NeurIPS (2021)
2. Andreini, P., Bonechi, S., Dimitri, G.M.: Enhancing glomeruli segmentation through cross-species pre-training. Neurocomputing **563** (2024). https://doi.org/10.1016/j.neucom.2023.126947, https://www.scopus.com/inward/record.uri?eid=2-s2.0-85174834663&doi=10.1016%2fj.neucom.2023.126947&partnerID=40&md5=931692afd67e6e2951041f22336bbc30. cited by: 6; All Open Access, Green Open Access, Hybrid Gold Open Access
3. Bianchini, M., Andreini, P., Bonechi, S.: From pixels to diagnosis: AI-driven skin lesion recognition. In: Kwaśnicka, H., Jain, N., Markowska-Kaczmar, U., Lim, C.P., Jain, L.C. (eds.) Advances in Smart Healthcare Paradigms and Applications, Intelligent Systems Reference Library, vol. 244, pp. 115–135. Springer, Cham (2023). https://doi.org/10.1007/978-3-031-37306-0_6
4. Bonechi, S., Bianchini, M., Bongini, P., et al.: Fusion of visual and anamnestic data for the classification of skin lesions with deep learning. In: Cristani, M., Prati, A., Lanz, O., Messelodi, S., Sebe, N. (eds.) New Trends in Image Analysis and Processing, ICIAP 2019, Lecture Notes in Computer Science, vol. 11808. Springer, Cham (2019). https://doi.org/10.1007/978-3-030-30754-7_21
5. Bonechi, S.: ISIC_WSM: generating weak segmentation maps for the ISIC archive. Neurocomputing **523**, 69–80 (2023)
6. Bonechi, S.: Leveraging segmentation maps to improve skin lesion classification. In: ESANN 2025 - Proceedings, 33th European Symposium on Artificial Neural Networks, Computational Intelligence and Machine Learning (2025)
7. Brinker, T.J., Hekler, A., Enk, A.H., et al.: Deep neural networks are superior to dermatologists in melanoma image classification. Eur. J. Cancer **119**, 11–17 (2019). https://doi.org/10.1016/j.ejca.2019.05.023
8. Caron, M., et al.: Emerging properties in self-supervised vision transformers. In: Proceedings of the IEEE/CVF International Conference on Computer Vision, pp. 9650–9660 (2021)
9. Cassidy, B.: Analysis of the ISIC image datasets: usage, benchmarks and recommendations, vol. 75, p. 102305 (2022)
10. Christopoulos, D., Spanos, S., Baltzi, E., Ntouskos, V., Karantzalos, K.: Skin lesion phenotyping via nested multi-modal contrastive learning. In: CVPR Workshop on Medical Vision (2025)
11. Dimitri, G.M., et al.: Deep learning approaches for the segmentation of glomeruli in kidney histopathological images. Mathematics **10**(11) (2022).https://

doi.org/10.3390/math10111934, https://www.scopus.com/inward/record.uri?eid=2-s2.0-85132429044&doi=10.3390%2fmath10111934&partnerID=40&md5=ddb5153abea79dcd84c8ddb4b39434da. cited by: 14; All Open Access, Gold Open Access, Green Open Access
12. Dosovitskiy, A., et al.: An image is worth 16x16 words: transformers for image recognition at scale. arXiv preprint: arXiv:2010.11929 (2020)
13. Du, S., et al.: Boosting dermatoscopic lesion segmentation via diffusion models with visual and textual prompts. arXiv preprint: arXiv:2310.02906 (2023)
14. Faghihi, A., Fathollahi, M., Rajabi, R.: Diagnosis of skin cancer using VGG16 and VGG19 based transfer learning models. arXiv 2404.01160 (2024). arXiv:2404.01160v1
15. Flosdorf, C., Engelker, J., Keller, I., Mohr, N.: Skin cancer detection utilizing deep learning: Classification of skin lesion images using a vision transformer. arXiv **2407.18554** (2024). arXiv:2407.18554v2
16. Goldblum, M., et al.: Battle of the backbones: a large-scale comparison of pretrained models across computer vision tasks. In: Advances in Neural Information Processing Systems, vol. 36, pp. 29343–29371 (2023)
17. Hao, Y., Xu, J., Wang, J.: SDSeg: diffusion model is a new paradigm for medical image segmentation. arXiv preprint: arXiv:2306.17118 (2023)
18. Himel, S., Rahman, R., Ahmed, S., Haque, E., Lee, S.W.: Lesion segmentation enhanced vision transformers for skin lesion classification. IEEE Access **12**, 20012–20025 (2024)
19. Ho, J., Chan, W., Saharia, C., Fleet, D.J., Norouzi, M., Salimans, T.: Classifier-free diffusion guidance (2022)
20. Ho, J., Jain, A., Abbeel, P.: Denoising diffusion probabilistic models. In: Advances in Neural Information Processing Systems (NeurIPS), vol. 33, pp. 6840–6851 (2020)
21. Islam, M.N., Hasan, M.R., Uddin, M.S.: SkinDistilViT: lightweight vision transformer for skin lesion classification. In: International Conference on Machine Learning and Applications (ICMLA) (2023)
22. Jin, Q., et al.: Cascade knowledge diffusion network for skin lesion diagnosis and segmentation. Appl. Soft Comput. **113**, 106881 (2021)
23. Karki, S., Devine, C., Lin, Z., Shen, D.: Derm-T2IM: text-to-image generation via diffusion models for skin lesion classification. IEEE J. Biomed. Health Inform. (2024). https://doi.org/10.1109/JBHI.2024.10781852
24. Khan, M.A., Mazhar, T., Ali, M.D., et al.: Automatic melanoma and non-melanoma skin cancer diagnosis using advanced adaptive fine-tuned convolution neural networks. Discov. Oncol. **16**(645) (2025).https://doi.org/10.1007/s12672-025-02279-8
25. Kirillov, A., et al.: Segment anything. In: 2023 IEEE/CVF International Conference on Computer Vision (ICCV), pp. 3992–4003 (2023). https://api.semanticscholar.org/CorpusID:257952310
26. Launer, R., Joskowicz, L., Greenspan, H.: CLIP-Lab: contrastive language–image pre-training for medical image retrieval and exploration. Med. Image Anal. **87**, 102839 (2023)
27. Liu, Z., et al.: Swin transformer: hierarchical vision transformer using shifted windows. In: Proceedings of the IEEE/CVF International Conference on Computer Vision (ICCV), pp. 10012–10022 (2021)
28. Mahbod, A.: PanDerm: a multimodal vision foundation model for dermatology. Nat. Med. (2024). in press

29. Naeem, A., Anees, T., Fiza, M., Naqvi, R.A., Lee, S.: SCDNet: a deep learning-based framework for the multiclassification of skin cancer using dermoscopy images. Sensors **22**(15), 5652 (2022). https://doi.org/10.3390/s22155652
30. Novikov, A.A., Lenis, D., Major, D., Hladvka, J., Wimmer, M., Bühler, K.: Fully convolutional architectures for multiclass segmentation in chest radiographs. IEEE Trans. Med. Imaging **37**(8), 1865–1876 (2018)
31. Radford, A., et al.: Learning transferable visual models from natural language supervision. In: Meila, M., Zhang, T. (eds.) Proceedings of the 38th International Conference on Machine Learning. Proceedings of Machine Learning Research, vol. 139, pp. 8748–8763. PMLR (2021). https://proceedings.mlr.press/v139/radford21a.html
32. Radford, A., et al.: Learning transferable visual models from natural language supervision (2021). arXiv preprint
33. Rombach, R., Blattmann, A., Lorenz, D., Esser, P., Ommer, B.: High-resolution image synthesis with latent diffusion models. In: Proceedings of the IEEE/CVF Conference on Computer Vision and Pattern Recognition (CVPR), pp. 10684–10695 (2022)
34. Ronneberger, O., Fischer, P., Brox, T.: U-Net: convolutional networks for biomedical image segmentation. In: Medical Image Computing and Computer-Assisted Intervention–MICCAI 2015: 18th International Conference, Munich, Germany, 5–9 October 2015, Proceedings, Part III 18, pp. 234–241. Springer (2015)
35. Schuhmann, C., et al.: LAION-5B: an open large-scale dataset for training next generation image-text models. In: Advances in Neural Information Processing Systems, vol. 35, pp. 25278–25294 (2022)
36. Shamshad, F., et al.: Transformers in medical imaging: a review. Med. Image Anal. **88**, 102802 (2023). https://doi.org/10.1016/j.media.2023.102802
37. Singh, M., Singh, A., Kukreja, V.: Generating synthetic images using stable diffusion model for skin lesion classification. In: 2024 5th International Conference on Electronics and Sustainable Communication Systems (ICESC), pp. 1593–1598 (2024). https://doi.org/10.1109/ICESC57846.2024.10499667
38. Sinha, S., Gupta, N.: Classification of melanocytic nevus images using BigTransfer (BiT): a study on a novel transfer learning–based method to classify melanocytic nevus images. In: 2022 International Conference on Computing, Communication, and Intelligent Systems (ICCCIS), pp. 708–712. IEEE (2022). https://doi.org/10.1109/ICCCIS56430.2022.10037627
39. Song, J., Meng, C., Ermon, S.: Denoising diffusion implicit models. arXiv preprint: arXiv:2010.02502 (2020)
40. Tan, M.: rethinking model scaling for convolutional neural networks. In: International Conference Machine Learning, vol. 97, pp. 6105–14 (2019). https://doi.org/10.48550.arXiv(1905)
41. Wang, R., et al.: GatedSegDiff: a gated fusion diffusion model for skin lesion segmentation. Med. Biol. Eng. Comput., 1–14 (2025)
42. Wang, Y., Tang, Y., Zhou, Y., Zhang, J., Zhang, Y., Shen, D.: BiomedCLIP: learning vision–language representations for biomedical applications. Nat. Mach. Intell. **5**, 500–510 (2023). https://doi.org/10.1038/s42256-023-00686-0
43. Yang, S., Shu, C., Hu, H., Ma, G., Yang, M.: Dermoscopic image classification of pigmented nevus under deep learning and the correlation with pathological features. Comput. Math. Methods Med., 9726181 (2022). https://doi.org/10.1155/2022/9726181

44. Yang, Y., Fu, H., Aviles-Rivero, A., Schönlieb, C.B., Zhu, L.: DiffMIC: dual-guidance diffusion network for medical image classification. arXiv preprint: arXiv:2303.10610 (2023)
45. You, K., et al.: CXR-CLIP: toward large scale chest x-ray language-image pre-training (2023)
46. Zhang, X., Xu, M., Qiu, D., Yan, R., Lang, N., Zhou, X.: MediCLIP: adapting CLIP for few-shot medical image anomaly detection. In: International Conference on Medical Image Computing and Computer-Assisted Intervention, pp. 458–468. Springer (2024)
47. Zhang, Y., Duanmu, H., Wang, S., Liu, C., Zhang, P., Dou, Q.: Clip in medical imaging: Comparison study on generalization and explainability. arXiv preprint: arXiv:2301.09604 (2023)
48. Zhang, Y., Tang, Y., Lu, Y., Wu, X., Shen, D.: Contrastive learning of medical visual representations from paired images and text. Med. Image Anal. **80**, 102405 (2022). https://doi.org/10.1016/j.media.2022.102405
49. Zhao, Z., et al.: Clip in medical imaging: a survey. Med. Image Anal. (2025). https://doi.org/10.1016/j.media.2025.103551

# Comparing General and Domain-Specific Pre-trained Language Models for Natural Product Classification

Barbara Toniella Corradini[1(✉)], Monica Bianchini[1], Franco Scarselli[1], Lorenzo Nigi[1], Vittorio Sali[1], and Alessia Lucia Prete[1,2]

[1] Department of Information Engineering and Mathematics, University of Siena, Via Roma, 56, 53100 Siena, Italy
barbaratoniella.c@student.unisi.it

[2] Fondazione Toscana Life Sciences, Via Fiorentina, 1, 53100 Siena, Italy

**Abstract.** The classification of natural products (NPs) remains a central challenge in cheminformatics, due to their structural complexity and the limited availability of curated data. In this study, we explore whether general-purpose language models—pre-trained on non-chemical corpora—can achieve performance comparable to models explicitly trained on molecular data. We evaluate six models, including ChemBERTa-zinc-base- v1, ChemBERTa-77M-MLM, LLaMA, GPT-2, MPNet, and MiniLM, across two established datasets (REFMET and NPClassifier), using both unsupervised clustering (K-Means) and supervised classification (logistic regression, MLP). In addition, we assess model robustness under input degradation, introducing systematic SMILES occlusion and additive embedding noise.

Our results show that general-purpose models can match or even surpass domain-specific alternatives, not only in clean classification scenarios but also under perturbation. Notably, general-purpose models demonstrate strong resilience and generalization, underscoring the transferability of domain-agnostic architectures to molecular tasks. These findings support the use of general-purpose language models as reliable tools for natural product classification, highlighting their potential to address complex challenges in cheminformatics through large-scale pre-trained architectures. Project page: https://github.com/bcorrad/MetaboLM25.

**Keywords:** Secondary metabolites · Pre-trained language models · Large Language Models · Molecular embeddings · Cheminformatics

## 1 Introduction

Natural products (NPs) comprise a diverse and complex class of chemical compounds produced by living organisms. They are a primary source of pharmaceuticals, agrochemicals, and bioactive agents, and play a central role in chemical ecology and drug discovery [2,14,19]. Automating the classification of natural


M. Bianchini et al. (Eds.): WIVACE 2025, CCIS 3001, pp. 61–74, 2027.
https://doi.org/10.1007/978-3-032-33185-4_5

products is a long-standing challenge in cheminformatics, due to their structural diversity, the sparsity of labelled data, and the continuous expansion of reference databases [3,12].

Traditional approaches to NP classification have relied on substructure analysis, rule-based systems, and handcrafted molecular descriptors [5,7]. These methods often fail to capture the full complexity and subtlety of natural product structures. The advent of deep learning and, in particular, sequence-based representations such as SMILES strings, has led to significant progress [11]. Sequence models, including convolutional neural networks and recurrent neural networks, have been used to generate molecular embeddings directly from SMILES, enabling better scalability and generalization [10].

Recently, transformer-based models have become the state-of-the-art in molecular representation learning. ChemBERTa, for example, adapts the BERT [8] architecture to large-scale SMILES data, showing strong results on molecular property prediction and transfer learning tasks [1,6]. Other models, such as SMILES Transformer [11] and Graphormer [23], have extended self-attention mechanisms to chemical graphs and sequences, achieving competitive performance in low-data regimes and on challenging property prediction benchmarks.

In parallel, the natural language processing community has produced a new generation of large language models (LLMs) trained on general text, including GPT-2 [15], LLaMA [21], and various sentence-transformer architectures [16,20,22]. These models, although not originally designed for chemistry, can ingest SMILES strings as input and generate powerful molecular embeddings [13,17]. Recent studies suggest that representations derived from LLMs like LLaMA and GPT-2 can match or even surpass domain-specific chemical models in molecular property prediction and virtual screening tasks [13,17]. Moreover, multi-modal and few-shot approaches leveraging LLMs have been explored for molecular annotation and reaction prediction [13].

Despite this rapid progress, robust evaluation of pre-trained language models for natural product classification remains limited. While methods like NPClassifier [12] and REFMET [3] have established taxonomies and gold-standard datasets, comprehensive studies comparing models pre-trained on general-purpose and chemical datasets for NP classification are lacking.

In this paper, we address these challenge through a systematic and comparative study. We extract molecular embeddings from a variety of pre-trained models, ranging from chemical-domain transformers (ChemBERTa) to large-scale general-purpose language models (LLaMA, GPT-2, MiniLM, MPNet, etc.), by passing SMILES strings as input. We then train downstream classifiers (logistic regression and multilayer perceptron) to predict NP classes, benchmarking all methods on the REFMET [3] and NPClassifier [12] datasets. We also assess embedding robustness by introducing controlled perturbations, including random noise and systematic occlusion, to the molecular representations and analysing performance across increasing perturbation levels.

Our contributions are as follows:

- We provide a comparison of six popular pre-trained chemical and non-chemical language models for natural product classification;
- We assess the robustness of different embeddings to noise and occlusions, providing insights into the generalization capabilities of general-purpose versus chemistry-specific models;
- We demonstrate that LLaMA-based and ChemBERTa-based embeddings both offer strong performance, and that architectural choices can have a significant impact on the resilience of NP classification pipelines.

Our work combines chemical expertise with new techniques in AI language understanding. This combination helps to connect the fields of cheminformatics and artificial intelligence. It also improves the way we build reliable, precise, and adaptable systems for classifying natural products.

The paper is organised as follows: Sect. 2 reviews related work on natural product classification, molecular embeddings, and the application of large language models to chemistry. Section 3 describes the datasets, language models, and evaluation metrics used in our experiments. Section 5 presents the results, including classification performance and robustness analyses. Section 6 discusses the implications of our findings, and gives the conclusions.

## 2 Related Work

*Natural Product Classification* – Natural products (NPs) are structurally diverse and biologically relevant compounds, presenting substantial challenges for automated classification. Early computational approaches relied on substructure-based heuristics, rule-based systems, and handcrafted chemical descriptors [5,7]. These methods typically lacked scalability and failed to capture subtle structural variations in NP families. Recent neural models, such as NPClassifier [12], have demonstrated superior performance by training deep architectures on large molecular fingerprint databases. REFMET [3] complements this by providing a standardized taxonomy for metabolomics, which serves as a benchmark for evaluating classification models in chemical informatics.

*Molecular Embeddings from SMILES* – Simplified Molecular Input Line Entry Specification (SMILES) strings provide a compact textual representation of molecular structures, enabling their direct use with deep learning architectures. Initial works such as SMILES2Vec [10] used convolutional and recurrent models to learn chemical representations from SMILES. Subsequent advances adopted transformer architectures: ChemBERTa [6] demonstrated that self-supervised pre-training on massive SMILES corpora improves molecular property prediction across diverse downstream tasks. ChemBERTa-2 [1] scaled this approach further, incorporating larger datasets and model sizes. Alternative transformer-based methods such as SMILES Transformer [11] and Graphormer [23] have shown strong generalization, especially under low-data conditions.

*Large Language Models and Transfer to Chemistry* – Large Language Models (LLMs), including GPT-2 [15], GPT-3 [4], and LLaMA [21], have been applied to molecular SMILES despite being trained on non-chemical corpora. Recent work in [17] offers a systematic evaluation of LLaMA and GPT embeddings on chemical tasks. The study reveals that LLaMA-based SMILES embeddings consistently outperform those of GPT-2 and approach the performance of domain-specific models like ChemBERTa, both in molecular property prediction and drug-drug interaction classification. This work highlights the surprising transferability of general LLMs to molecular representation learning and motivates further exploration in chemical tasks involving structured data, such as NP classification. Similarly, ChemCrow [13] demonstrates how general LLMs can be enhanced with domain-specific tools to support tasks like synthesis planning, structure recognition, and property inference.

*Robustness and Perturbation Analysis* – Assessing model robustness under input degradation has become a priority in molecular machine learning. In [9], the stability of learned representations under distributional shifts and adversarial perturbations has been studied, while [18] extended this investigation by evaluating uncertainty and robustness in graph-based molecular models. However, most robustness analyses focus on chemical-domain models and do not account for embeddings derived from large-scale, non-chemical LLMs.

*Positioning Our Work* – We present a focused comparison of domain-specific and general-purpose language models for natural product classification, assessing their robustness in hierarchical metabolomics tasks. Although prior work shows that general-purpose models can extract chemical information from protein SMILES [17], it is unclear if this extends to the more diverse and complex natural product domain. We evaluate performance under random noise and systematic occlusion on two benchmark datasets using supervised classifiers (logistic regression, MLP) and an unsupervised approach (K-Means). Our results reveal how embedding choice affects classification reliability and offer practical guidance for robust molecular classification.

## 3 Materials and Methods

In this section, we describe the datasets used for our experiments (Sect. 3.1), the language models employed for embedding generation (Sect. 3.2), and the evaluation metrics adopted (Sect. 3.3).

### 3.1 Datasets

**NPClassifier.** The NPClassifier dataset [12] consists of 73,607 natural product structures aggregated from major public repositories, including PubChem, ChEBI, ChemSpider, and UNPD. Each molecule is annotated with a hierarchical, three-level taxonomy: Pathway (7 categories), Superclass (70 categories),

**Table 1.** Number of original, filtered and ablation samples with classes for each dataset and NPClassifier hierarchy.

| Dataset | Original | Filtered | Ablation | Classes |
|---|---|---|---|---|
| NPClassifier superclass | 77,707 | 76,463 | 11,350 | 70 |
| NPClassifier pathway | 77,707 | 74,223 | 1,400 | 7 |
| NPClassifier class | 77,707 | 73,706 | 44,788 | 609 |
| REFMET | 187,745 | 29,684 | 3,170 | 20 |

and Class (672 categories). Structural representations are derived from counted Morgan fingerprints, capturing the frequency of substructural motifs rather than their binary presence. The taxonomy was systematically constructed from literature in specialized metabolism and validated against the MIBiG database to ensure biosynthetic relevance.

**REFMET.** The REFMET dataset [3] is derived form 280,000 named metabolites from more than 1,400 mass spectrometry and NMR studies archived in the National Metabolomics Data Repository. It features a four-level hierarchical annotation, spanning from detailed molecular structure to broad metabolite classes. In this work, we focus on the Superclass level (20 categories), which offers an optimal balance between semantic resolution and computational tractability, while aligning with prevalent categorizations in metabolomics research.[1] Since the REFMET dataset does not directly include SMILES representations, we programmatically retrieved them via the PubChem PUG REST API using the provided PubChem Compound IDs (CIDs). Specifically, we queried batches of CIDs to obtain their corresponding Isomeric SMILES, and integrated the results back into the dataset.[2] The process included CID validation, batched requests (to ensure API reliability), and assignment of SMILES strings to valid entries, with the resulting annotated dataset saved for downstream analysis.

**Data Selection.** For both datasets, we exclude label classes with fewer than five samples to ensure statistical reliability. In the NPClassifier dataset, all three annotation levels—Pathway, Superclass, and Class—are considered in our main experiments. For REFMET, we programmatically retrieved SMILES strings using PubChem CIDs and discarded entries without a valid SMILES representation. After filtering, we retained a total of 29,684 REFMET entries, and restricted our analysis to the Superclass level, which offers a balanced compromise between semantic specificity and class distribution.

[1] We retrieved the REFMET dataset from https://www.metabolomicsworkbench.org/databases/refmet/browse.php.

[2] We used the PubChem PUG REST API: https://pubchemdocs.ncbi.nlm.nih.gov/pug-rest.

To control for class imbalance in our ablation studies, we capped each remaining class to a maximum of 200 instances using uniform random sampling. Classes with fewer than 200 samples were retained in full. Table 1 reports the final distribution of classes and total samples, both before and after filtering and subsampling.

### 3.2 Language Models and Embedding Generation

To convert SMILES strings into fixed-size numerical embeddings, we employ six pre-trained language models spanning both domain-specific and general-purpose architectures. This diverse selection enables us to assess how model design and pre-training influence the quality of molecular representations.

**Chemistry-Specific Models.** In our evaluation, we include two ChemBERTa variants specifically trained on SMILES strings. ChemBERTa-77M-MLM is based on RoBERTa-base, a bidirectional transformer with 12 layers, 12 attention heads, and approximately 77 million parameters. It is trained using a Masked Language Modeling (MLM) objective on 10 million SMILES from PubChem. ChemBERTa-zinc-base-v1 adopts the same RoBERTa-base architecture but is pre-trained on a significantly larger corpus of 220 million molecules from the ZINC database, allowing it to capture a wider range of chemical substructures and relationships.

**General-Purpose Language Models.** To investigate the generalization of language models across domains, we include two autoregressive transformers. LLaMA3 is a state-of-the-art decoder-only model with billions of parameters (we use the 8B variant), featuring multi-query and grouped-query attention mechanisms to optimize inference and scalability. It is trained on a large, mixed-domain corpus including code and natural language. GPT-2, with approximately 117M parameters in its base configuration, consists of 12 transformer decoder layers and was trained on the WebText dataset. While outdated compared to recent models, it serves as a strong historical baseline for general-purpose modeling.

**Sentence Embedding Models.** We also evaluate two models from the Sentence-Transformers library, designed to produce semantically meaningful sentence embeddings. all-mpnet-base-v2 is based on Microsoft MPNet architecture, combining bidirectional attention with permutation-based pre-training; it contains 12 layers and 110M parameters. all-MiniLM-L6-v2 is a distilled version of a transformer model with only 6 layers and around 22M parameters, offering a high-efficiency option for fast embedding generation without substantial performance loss.

**Embedding Extraction.** For all models, we extract a fixed-dimensional embedding for each SMILES string by applying mean pooling over the token embeddings from the final hidden layer. This results in a dense vector that summarizes the global semantics of the molecular structure as captured by the model.

### 3.3 Evaluation Metrics

We evaluate the embedding quality via both supervised classification and unsupervised clustering metrics.

**Supervised Classification.** Given true positives (TPs), false positives (FPs), false negatives (FNs), and true negatives (TNs), we report the following macro-averaged metrics over all classes:

- **Precision** measures the accuracy of positive predictions, indicating the proportion of correctly predicted positive instances out of all instances predicted as positive:
$$\text{Precision} = \frac{\text{TP}}{\text{TP} + \text{FP}}$$
- **Recall** indicates the proportion of correctly predicted positive instances out of all actual positive instances:
$$\text{Recall} = \frac{\text{TP}}{\text{TP} + \text{FN}}$$
- **F1 Score** is the harmonic mean of precision and recall:
$$\text{F1} = \frac{2 \cdot \text{TP}}{2 \cdot \text{TP} + \text{FP} + \text{FN}}$$
- **Accuracy** is the proportion of correct predictions over all samples:
$$\text{Accuracy} = \frac{\text{TP} + \text{TN}}{\text{TP} + \text{FP} + \text{FN} + \text{TN}}$$

**Unsupervised Clustering.** We apply K-Means to the embeddings and compare the resulting clusters to ground-truth labels:

- **Normalized Mutual Information (NMI)** quantifies the shared information between predicted and true labels, normalized by their entropy:
$$\text{NMI}(U, V) = \frac{2 \cdot I(U; V)}{H(U) + H(V)}$$
- **Adjusted Rand Index (ARI)** measures similarity between clustering and ground truth, corrected for chance:
$$\text{ARI} = \frac{\text{RI} - \mathbb{E}[\text{RI}]}{\max(\text{RI}) - \mathbb{E}[\text{RI}]}$$

## 4 Experiments

We conducted a comprehensive evaluation of molecular embeddings derived from six pre-trained language models across multiple classification tasks and perturbation scenarios.

*Evaluation Pipeline* – For each model, we computed the SMILES embeddings of each molecule. These embeddings were then evaluated using three classifiers: K-Means for unsupervised clustering, and two supervised models—logistic regression and a multilayer perceptron (MLP)—both implemented in PyTorch. Our experiments were performed on a machine with an NVIDIA GeForce RTX 4090 GPU and 128GB RAM. Each model was trained and evaluated using stratified 60/40 training-test splits, with performance assessed using macro-averaged accuracy, F1 score, precision, recall, ARI, and NMI.

*Dataset Processing* – To ensure statistical reliability, we excluded rare classes with fewer than five samples. For robustness analyses, we applied stratified subsampling to cap each class at 200 samples.

*Robustness Studies* – To assess generalization under input degradation, we introduced two types of perturbations:

1. **SMILES occlusion**, where increasing fractions of 10%, 30%, 50% and 70% characters in each SMILES string were randomly masked;
2. **Embedding noise**, where Gaussian noise of varying standard deviation of 0.1, 0.3, 0.5 and 0.7 was injected directly into the frozen embeddings.

In both cases, new training-test splits were generated, and all classifiers were re-evaluated at each perturbation level. Embeddings for occluded SMILES were re-computed in batches on GPU using model-specific embedding functions. Noise was applied directly in NumPy space for computational efficiency.

**Table 2.** NPClassifier and REFMET dataset classification results.

(a) REFMET

| Model | Classifier | ARI ↑ | NMI ↑ | Acc ↑ | F1 ↑ | Prec ↑ | Recall ↑ |
|---|---|---|---|---|---|---|---|
| | KMeans | 0.329 | 0.554 | – | – | – | – |
| ChemBERTa-77 | LR | – | – | 0.920 | 0.741 | 0.773 | 0.732 |
| (Chem) | MLP | – | – | 0.928 | 0.743 | 0.755 | 0.738 |
| | KMeans | 0.304 | 0.487 | – | – | – | – |
| ChemBERTa-zinc | LR | – | – | 0.905 | 0.675 | 0.718 | 0.662 |
| (Chem) | MLP | – | – | 0.907 | 0.660 | 0.690 | 0.648 |
| | KMeans | 0.082 | 0.206 | – | – | – | – |
| GPT | LR | – | – | 0.907 | 0.710 | 0.757 | 0.696 |
| (GP) | MLP | – | – | 0.911 | 0.677 | 0.735 | 0.668 |
| | KMeans | 0.339 | 0.542 | – | – | – | – |
| LLaMA | LR | – | – | 0.894 | 0.670 | 0.770 | 0.637 |
| (GP) | MLP | – | – | 0.914 | 0.704 | 0.772 | 0.688 |
| | KMeans | 0.345 | 0.506 | – | – | – | – |
| MPNet | LR | – | – | 0.902 | 0.726 | 0.774 | 0.711 |
| (GP) | MLP | – | – | 0.920 | 0.736 | 0.798 | 0.717 |
| | KMeans | 0.281 | 0.505 | – | – | – | – |
| MiniLM | LR | – | – | 0.898 | 0.695 | 0.755 | 0.666 |
| (GP) | MLP | – | – | 0.925 | 0.753 | 0.817 | 0.724 |

(b) NPClassifier – Pathway

| Model | Classifier | ARI ↑ | NMI ↑ | Acc ↑ | F1 ↑ | Prec ↑ | Recall ↑ |
|---|---|---|---|---|---|---|---|
| | KMeans | 0.452 | 0.508 | – | – | – | – |
| ChemBERTa-77 | LR | – | – | 0.948 | 0.926 | 0.931 | 0.921 |
| (Chem) | MLP | – | – | 0.969 | 0.954 | 0.949 | 0.959 |
| | KMeans | 0.310 | 0.375 | – | – | – | – |
| ChemBERTa-zinc | LR | – | – | 0.924 | 0.887 | 0.901 | 0.875 |
| (Chem) | MLP | – | – | 0.949 | 0.920 | 0.926 | 0.916 |
| | KMeans | 0.060 | 0.100 | – | – | – | – |
| GPT | LR | – | – | 0.931 | 0.899 | 0.906 | 0.894 |
| (GP) | MLP | – | – | 0.943 | 0.914 | 0.917 | 0.911 |
| | KMeans | 0.440 | 0.417 | – | – | – | – |
| LLaMA | LR | – | – | 0.951 | 0.916 | 0.932 | 0.909 |
| (GP) | MLP | – | – | 0.947 | 0.921 | 0.927 | 0.917 |
| | KMeans | 0.230 | 0.300 | – | – | – | – |
| MPNet | LR | – | – | 0.888 | 0.836 | 0.856 | 0.821 |
| (GP) | MLP | – | – | 0.942 | 0.915 | 0.915 | 0.915 |
| | KMeans | 0.240 | 0.317 | – | – | – | – |
| MiniLM | LR | – | – | 0.880 | 0.827 | 0.843 | 0.814 |
| (GP) | MLP | – | – | 0.942 | 0.918 | 0.921 | 0.917 |

(c) NPClassifier – Superclass

| Model | Classifier | ARI ↑ | NMI ↑ | Acc ↑ | F1 ↑ | Prec ↑ | Recall ↑ |
|---|---|---|---|---|---|---|---|
| | KMeans | 0.249 | 0.542 | – | – | – | – |
| ChemBERTa-77 | LR | – | – | 0.888 | 0.815 | 0.849 | 0.796 |
| (Chem) | MLP | – | – | 0.929 | 0.865 | 0.894 | 0.848 |
| | KMeans | 0.185 | 0.449 | – | – | – | – |
| ChemBERTa-zinc | LR | – | – | 0.848 | 0.747 | 0.791 | 0.731 |
| (Chem) | MLP | – | – | 0.864 | 0.739 | 0.801 | 0.713 |
| | KMeans | 0.078 | 0.277 | – | – | – | – |
| GPT | LR | – | – | 0.838 | 0.741 | 0.800 | 0.724 |
| (GP) | MLP | – | – | 0.848 | 0.701 | 0.766 | 0.671 |
| | KMeans | 0.208 | 0.463 | – | – | – | – |
| LLaMA | LR | – | – | 0.837 | 0.711 | 0.818 | 0.700 |
| (GP) | MLP | – | – | 0.817 | 0.577 | 0.687 | 0.554 |
| | KMeans | 0.175 | 0.399 | – | – | – | – |
| MPNet | LR | – | – | 0.750 | 0.608 | 0.692 | 0.574 |
| (GP) | MLP | – | – | 0.851 | 0.730 | 0.791 | 0.698 |
| | KMeans | 0.176 | 0.425 | – | – | – | – |
| MiniLM | LR | – | – | 0.736 | 0.579 | 0.669 | 0.542 |
| (GP) | MLP | – | – | 0.857 | 0.753 | 0.799 | 0.731 |

(d) NPClassifier – Class

| Model | Classifier | ARI ↑ | NMI ↑ | Acc ↑ | F1 ↑ | Prec ↑ | Recall ↑ |
|---|---|---|---|---|---|---|---|
| | KMeans | 0.169 | 0.668 | – | – | – | – |
| ChemBERTa-77 | LR | – | – | 0.862 | 0.756 | 0.810 | 0.736 |
| (Chem) | MLP | – | – | 0.859 | 0.742 | 0.791 | 0.725 |
| | KMeans | 0.121 | 0.615 | – | – | – | – |
| ChemBERTa-zinc | LR | – | – | 0.798 | 0.672 | 0.750 | 0.652 |
| (Chem) | MLP | – | – | 0.773 | 0.619 | 0.688 | 0.602 |
| | KMeans | 0.035 | 0.446 | – | – | – | – |
| GPT | LR | – | – | 0.777 | 0.623 | 0.729 | 0.603 |
| (GP) | MLP | – | – | 0.736 | 0.563 | 0.638 | 0.548 |
| | KMeans | 0.130 | 0.619 | – | – | – | – |
| LLaMA | LR | – | – | 0.735 | 0.556 | 0.684 | 0.547 |
| (GP) | MLP | – | – | 0.588 | 0.234 | 0.291 | 0.240 |
| | KMeans | 0.126 | 0.591 | – | – | – | – |
| MPNet | LR | – | – | 0.708 | 0.523 | 0.599 | 0.503 |
| (GP) | MLP | – | – | 0.756 | 0.604 | 0.669 | 0.580 |
| | KMeans | 0.137 | 0.605 | – | – | – | – |
| MiniLM | LR | – | – | 0.690 | 0.480 | 0.571 | 0.460 |
| (GP) | MLP | – | – | 0.769 | 0.617 | 0.678 | 0.598 |

## 5 Results

Table 2 and Figs. 1 and 2 summarize the performance of all models across datasets and experimental conditions.

*Overall, our findings confirm that general-purpose language models can match, and in some cases rival, domain-specific models for metabolite classification—both in standard settings and under perturbations.*

In clean conditions, the best overall results are obtained by ChemBERTa-77 and LLaMA, which consistently rank at the top across datasets and classification levels. On REFMET, ChemBERTa-77 paired with an MLP achieves the highest accuracy (0.928) and F1 score (0.743). MiniLM and MPNet achieve comparable performance (0.925 accuracy and 0.753 F1 for MiniLM) despite the absence of chemistry-specific pre-training, even surpassing ChemBERTa-zinc. Similar trends are observed on NPClassifier, where general-purpose models achieve up

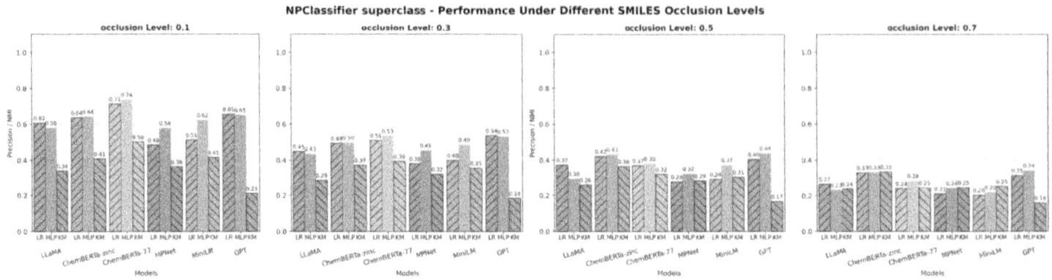


(a) NPClassifier superclass

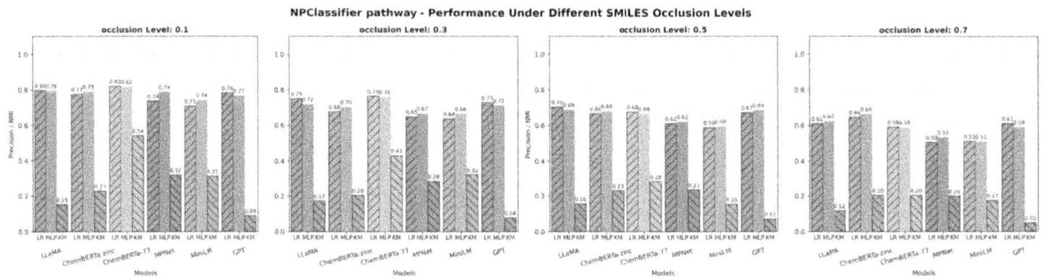


(b) NPClassifier pathway

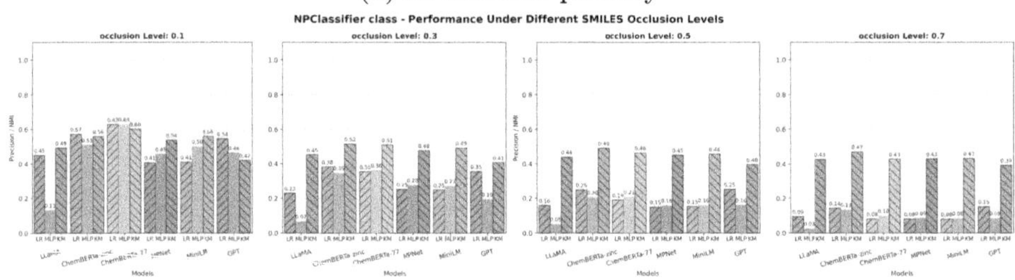


(c) NPClassifier class

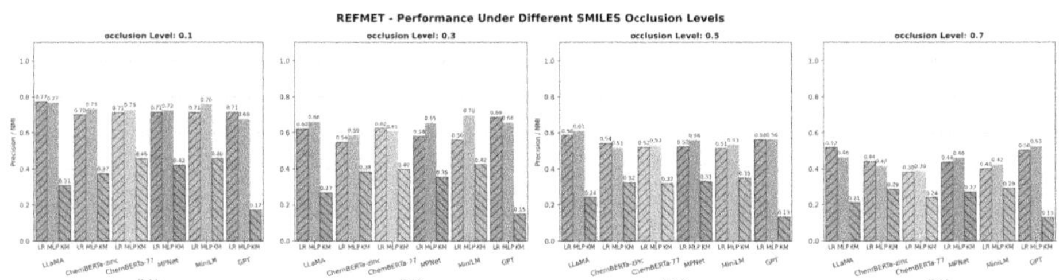


(d) REFMET

**Fig. 1.** Model accuracy under increasing SMILES occlusion for each dataset and NPClassifier hierarchy. All models experience a decline in performance as more of the input is masked, but LLaMA and ChemBERTa remain consistently more robust across all settings.

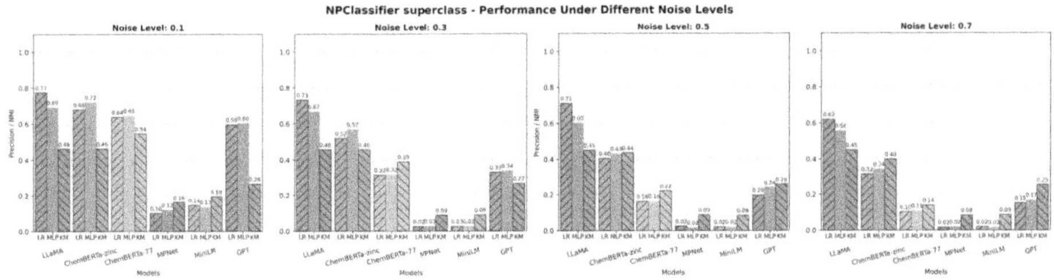


(a) NPClassifier superclass

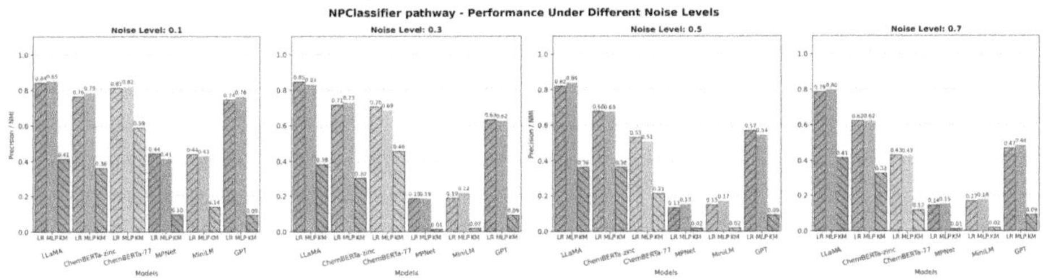


(b) NPClassifier pathway

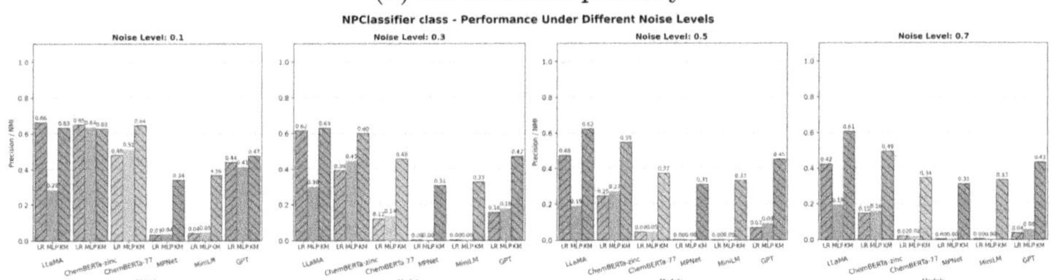


(c) NPClassifier class

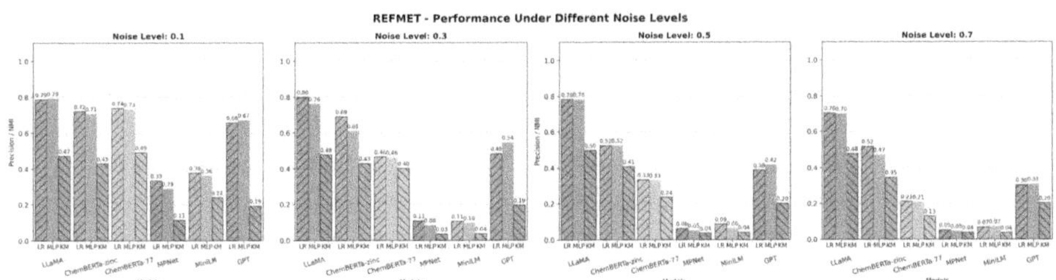


(d) REFMET

**Fig. 2.** Model accuracy under increasing label noise for each dataset and NPClassifier hierarchy. All models show a decline in performance as noise increases, but LLaMA and ChemBERTa consistently demonstrate greater robustness across all settings.

to 0.943 accuracy at the pathway level. They remain competitive even at the finest levels of superclass and class.

K-Means clustering confirms that these models capture latent class structure. ARI and NMI scores show that ChemBERTa-77, LLaMA, and MPNet produce embeddings aligned with ground-truth labels, reaching ARI values above 0.33 on REFMET.

Robustness is evaluated under two perturbation settings: SMILES occlusion and embedding noise. Graphical results on the accuracy of K-Means, logistic regression, and MLP are reported in Figs. 2 and 1; further details are available on the project page. In the occlusion experiments (Fig. 1), models were tested under progressive masking of contiguous SMILES fragments. While all models decline in accuracy with increased occlusion, ChemBERTa and LLaMA maintain superior performance, particularly at the pathway and superclass levels, suggesting they capture global chemical features. Degradation is more pronounced at the class level and on REFMET, but these models still outperform the others.

The noise experiments (Fig. 2) reveal a similar pattern. As Gaussian noise is added to the embeddings, all models show reduced performance. However, ChemBERTa and LLaMA degrade more gracefully. Their resilience becomes increasingly important in fine-grained tasks, where label noise and class imbalance have greater impact.

In summary, general-purpose language models (especially LLaMA, MPNet, and MiniLM) generate molecular embeddings that are both effective and robust. They perform competitively with domain-specific models and retain accuracy under perturbation. While ChemBERTa remains the top-performing model overall, the small performance gap suggests that general-purpose architectures are viable and reliable alternatives, even without chemical-domain pre-training.

## 6 Conclusions

In this study, we systematically evaluated the effectiveness of domain-specific and general-purpose language models for metabolite classification, using both REFMET and NPClassifier datasets. We compared embeddings derived from six pre-trained models across multiple classification levels, supervised classifiers, and robustness scenarios.

Our results prove that general-purpose models, such as LLaMA, MiniLM, and MPNet, perform on par with domain-specific models like ChemBERTa-77. They achieve high accuracy and F1 scores in clean settings and exhibit strong resilience to input degradation. Even in challenging classification tasks—such as those involving incomplete SMILES strings or noisy embeddings—general-purpose models retain competitive performance. This suggests that large-scale language modeling on non-chemical corpora still yields embeddings that are chemically informative and transferable.

Although ChemBERTa-77 remains the most consistent top performer, the marginal differences observed across most experiments highlight that domain-specific pre-training is not strictly necessary to achieve state-of-the-art results in metabolite classification. The robustness of general-purpose models under occlusion and noise further reinforces their suitability for real-world applications, where data may be noisy, sparse, or partially observed.

Moreover, the inherent flexibility of general-purpose architectures allows seamless adaptation to new tasks and domains without requiring expensive domain-specific retraining. This flexibility, combined with their competitive performance, makes them valuable assets in metabolomics and cheminformatics pipelines, especially in scenarios with limited labeled data or domain expertise.

In summary, our findings support the use of general-purpose pre-trained models as reliable, efficient, and adaptable tools for metabolite analysis. This opens new opportunities for deploying scalable, domain-agnostic NLP architectures in cheminformatics workflows.

## References


1. Ahmad, Y., Smith, J., Kim, S.: ChemBERTa-2: larger-scale self-supervised pre-training for chemical and biological molecules. arXiv preprint: arXiv:2109.04813 (2021)
2. Atanasov, A.G., Zotchev, S.B., Dirsch, V.M., Supuran, C.T.: Natural products in drug discovery: advances and opportunities. Nat. Rev. Drug Discovery **20**(3), 200–216 (2021). https://doi.org/10.1038/s41573-020-00114-z
3. Blum, L.C., et al.: RefMet 2.0: an updated reference taxonomy and database for metabolomics. Metabolomics **17**(9), 92 (2021). https://doi.org/10.1007/s11306-021-01850-0
4. Brown, T.B., et al.: Language models are few-shot learners. arXiv preprint: arXiv:2005.14165 (2020)
5. Chevrette, M.G., Currie, C.R.: The value and distribution of secondary metabolism in bacterial genomes. Proc. Natl. Acad. Sci. **118**(43), e2110877118 (2021). https://doi.org/10.1073/pnas.2110877118
6. Chithrananda, S., Grand, G., Ramsundar, B.: ChemBERTa: large-scale self-supervised pretraining for molecular property prediction. arXiv preprint: arXiv:2010.09885 (2020)
7. Christofidou, E., Li, X., Woodcock, S.: Natural product–derived libraries for drug discovery. Drug Discov. Today **25**(11), 2003–2010 (2020). https://doi.org/10.1016/j.drudis.2020.07.005
8. Devlin, J., Chang, M.W., Lee, K., Toutanova, K.: BERT: pre-training of deep bidirectional transformers for language understanding. In: Proceedings of the 2019 Conference of the North American Chapter of the Association for Computational Linguistics: Human Language Technologies, vol. 1 (long and short papers), pp. 4171–4186 (2019)
9. Feinberg, E.N., Joshi, E., Pande, V.S., Cheng, A.C.: Improvement in ADMET prediction with multitask deep featurization. J. Med. Chem. **63**(16), 8835–8848 (2020)
10. Goh, G.B., Hodas, N.O., Siegel, C., Vishnu, A.: SMILES2vec: an interpretable general-purpose deep neural network for predicting chemical properties. arXiv preprint: arXiv:1712.02034 (2017)
11. Honda, S., Shi, S., Ueda, H.R.: Smiles transformer: Pre-trained molecular fingerprint for low data drug discovery. arXiv preprint: arXiv:1911.04738 (2019)

12. Kim, S., Cummings, M.D., Yevshin, I.: NPClassifier: a deep neural network-based structural classification tool for natural products. J. Nat. Prod. **84**(2), 298–306 (2021). https://doi.org/10.1021/acs.jnatprod.1c00399
13. Malla, Y., Gao, K., Veličković, P., Coley, C.W.: ChemCrow: enhancing LLMs with domain-specific tools for chemistry. In: Conference on AI in Chemistry (2023). to appear
14. Newman, D.J., Cragg, G.M.: Natural products as sources of new drugs from 1981 to 2014. J. Nat. Prod. **79**(3), 629–661 (2016). https://doi.org/10.1021/acs.jnatprod.5b01055
15. Radford, A., Wu, J., Child, R., Luan, D., Amodei, D., Sutskever, I.: Language models are unsupervised multitask learners. OpenAI Blog (2019)
16. Reimers, N., Gurevych, I.: Sentence-BERT: sentence embeddings using Siamese BERT-networks. arXiv preprint: arXiv:1908.10084 (2019)
17. Sadeghi, M., Wolf, T., Wattanachai, N., Lehrig, S.: Simulation not available; systematic evaluation of llama and GPT embeddings on chemical tasks. BMC Bioinformatics **25**(1), 123 (2024). https://doi.org/10.1186/s12859-024-04567-8
18. Schwalbe-Koda, D., Tan, A.R., Gómez-Bombarelli, R.: Differentiable sampling of molecular geometries with uncertainty-based adversarial attacks. Nat. Commun. **12**(1), 5104 (2021)
19. da Silva, L., Hauser, N., Taglialatela-Scafati, O.: Review on natural products databases: where to find data in 2020. J. Cheminform. **11**, 58 (2019). https://doi.org/10.1186/s13321-019-0384-0
20. Song, K., Tan, X., Qin, T., Lu, J., Liu, T.Y.: MPNet: masked and permuted pre-training for language understanding. In: Advances in Neural Information Processing Systems, vol. 33, pp. 16857–16867 (2020)
21. Touvron, H.: LLaMA: open and efficient foundation language models. arXiv preprint: arXiv:2302.13971 (2023)
22. Wang, W., Wei, F., Dong, L., Bao, H., Yang, N., Zhou, M.: MiniLM: deep self-attention distillation for task-agnostic compression of pre-trained transformers. In: Larochelle, H., Ranzato, M., Hadsell, R., Balcan, M., Lin, H. (eds.) Advances in Neural Information Processing Systems, vol. 33, pp. 5776–5788. Curran Associates, Inc. (2020). https://proceedings.neurips.cc/paper_files/paper/2020/file/3f5ee243547dee91fbd053c1c4a845aa-Paper.pdf
23. Ying, R., et al.: Graphormer: A transformer-based graph neural network for molecular property prediction. arXiv preprint: arXiv:2106.05234 (2021)

# Electrical Dynamics in Living *Scenedesmus Acutus*

Salvatore Del Basso[1,3], Alba Amato[2], Pietro Squeri[1], Anna Torelli[3], Matteo Marieschi[3], Roberta Ruotolo[3], Saksham Sharma[4], Neil Phillips[4], Andrew Adamatzky[4], Pasquale D'Angelo[1], Riccardo Pecori[1,5(✉)], and Giuseppe Tarabella[1]

[1] Institute of Materials for Electronics and Magnetism, National Research Council of Italy, Parma, PR, Italy
{riccardo.pecori,giuseppe.tarabella}@cnr.it
[2] Department of Political Science, University of Campania "Luigi Vanvitelli", Caserta, CE, Italy
[3] Department of Chemistry, Life Sciences and Environmental Sustainability, University of Parma, Parma, PR, Italy
[4] Unconventional Computing Lab, School of Computing and Creative Technologies, University of the West of England, Bristol, UK
[5] Department of Applied and Theoretical Sciences, eCampus University, Novedrate, CO, Italy

**Abstract.** In this paper, we analyzed the spontaneous and stimulated electrical activity of *Scenedesmus acutus* algae, a green, non-motile, unicellular, freshwater photosynthetic microalga, belonging to the order of Sphaeropleales. The results, obtained by using computational techniques such as Fast Fourier Transform, Periodogram, Spectrogram, and LSTM-based Autoencoders, demonstrate a spontaneous electrical activity, highlighting some low frequencies whose spectral power is augmented in presence of a proper electrical stimulus applied to the solution containing the algae. The same frequency patterns are not evident or only minimal when considering the medium alone. These preliminary outcomes demonstrate the chance of using *S. acutus* as a possible building block for future bioelectronic computing systems.

**Keywords:** Living circuits · *Scenedesmus acutus* · Pattern Recognition · Algae · Bioelectronics · Signal Dynamics · FFT · LSTM · Autoencoder

## 1 Introduction

Organisms lacking a nervous system, such as bacteria, slime molds, fungi, and plants, exhibit electrical activity analogous to neuronal spiking [1]. Recording these signals offers critical insights into the organism's physiological status. Moreover, the analysis of this electrical behavior facilitates the investigation of responses to stressors or external stimuli (e.g., electrical, mechanical, chemical, or optical). The capacity to control these biological responses presents novel opportunities for integrating living systems into electronic, sensing, memory, or hybrid circuits, thus enabling the development of


M. Bianchini et al. (Eds.): WIVACE 2025, CCIS 3001, pp. 75–85, 2027.
https://doi.org/10.1007/978-3-032-33185-4_6

prototypical "living" devices. Integrating living organisms into electrical sensing platforms offers a cost-effective, environmentally benign, and robust solution for continuous monitoring and sensing applications [2]. Recent studies are evaluating the suitability of sensing devices that utilize living algae as the core component for biosensing applications [3, 4].

Herein, we investigated both the spontaneous and electrically stimulated activity of *S. acutus* with the overarching goal of engineering functional biological electrical circuits. We employed signal processing and pattern recognition techniques, mainly based on Fourier Transform and Autoencoder Neural Networks, to analyze the experimental data. The analysis identified low-frequency peaks in the voltage signals, which were markedly enhanced when stimulated with external electrical stimuli at corresponding frequencies. These results underscore the potential of *S. acutus* as a fundamental building block for bioelectronic systems, fostering the long-term objective of this preliminary study, i.e., the realization of fully functional electrical circuits based on living algal structures.

The rest of the paper has the following structure. Section 2 summarizes the main information about the considered alga, i.e., *S. acutus*. Section 3 provides an overview of the considered experimental setup to cultivate the algae and to measure their electrical activity as well as of the considered computational analysis techniques. Section 4 shows some of the most relevant obtained results considering both *S. acutus* and its growing medium. Section 5 seals up the paper with some preliminary conclusions and provides future developments.

## 2 Background on *S. Acutus*

*Scenedesmus acutus* (*S. acutus*) [5] is a green, non-motile, unicellular, freshwater photosynthetic microalga, belonging to the order of Sphaeropleales (Fig. 1). It has a haplontic cycle and usually asexually reproduces through the production of autospores typically forming small colonies (coenobia) of 2, 4, or 8 spindle-shaped cells with acute apices. When the culture is kept under agitation, the colonies divide into single cells. This species is ubiquitous in freshwater environments, particularly in eutrophic (nutrient-rich) waters. As a key member of the phytoplankton community, it plays a vital role in aquatic ecosystems by contributing significantly to primary production. Due to its rapid growth rate and cultivation ease, *S. acutus* is extensively utilized in the fields of ecotoxicology and biotechnology, including applications in wastewater treatment, biofuel research, and ecotoxicological assays. In particular, it is used for i) bioremediation of water contaminated by heavy metals thanks to its ability to absorb and detoxify them due to its intrinsic sulfur metabolism; ii) the production of sustainable energy through bio-photovoltaic systems and photosynthetic fuel cells; iii) the development of biomaterials derived from microalgae, such as extracellular polysaccharides.

However, in this paper, our aim is to figure out the potential use of *S. acutus* in unconventional systems, exploiting its intrinsic electrical dynamic activity.

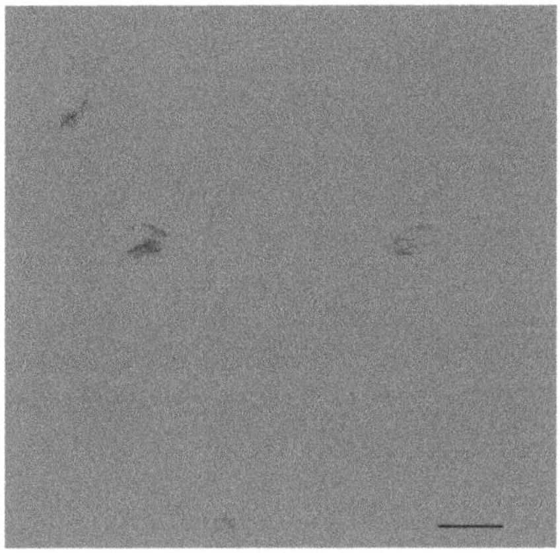

**Fig. 1.** Light microscope image of *Scenedesmus acutus.* It usually appears as flat colonies of 2, 4, or 8 cells arranged in rows, but can also appear as single cells. Scale bar = 10 μm.

# 3 Materials and Methods

## 3.1 In Vitro Culture of *S. Acutus*

Synchronized axenic cultures of *S. acutus* were maintained in liquid culture medium (US EPA, 1978 pH 7.7 ± 0.1, modified as described by Marieschi et al. in [6]), in a climate-controlled chamber (23 ± 1 °C, 230 μmol $m^{-2}$ $s^{-1}$ light intensity irradiance, white-cool fluorescent lamps, 16-h photoperiod). The cultures were continuously aerated by means of sterile filtered air.

We analyzed the voltage time-series generated by *S. acutus* cultures while they were monitored under two distinct experimental settings. In the first, the cultures were exposed to a controlled electrical stimulation (0.1 Hz and 10 mV), designed to perturb their native electrical behavior. In the second, the same cultures were observed in the absence of any external stimulation, allowing us to capture their spontaneous, baseline electrical activity. This comparison enabled us to distinguish intrinsic oscillatory features from those retrieved after the application of the aforementioned stimulus.

A multichannel Data Acquisition Unit (DAQ, Keithley 6510) was employed to continuously record the electrical signals generated by the *S. acutus* cultures. Pairs of monopolar, medical-grade needle electrodes (Spes Medica Spa) were connected to the DAQ and immersed in the algal suspension to capture local voltage changes with minimal perturbation of the system. The setup is shown in Fig. 2 and was housed inside a biological safety cabinet to maintain controlled and sterile conditions. A magnetic stir bar was placed in the beaker to provide a constant and low-intensity mixing of the alga, preventing aggregation, sedimentation, and maintaining a uniform suspension during long-term recordings.

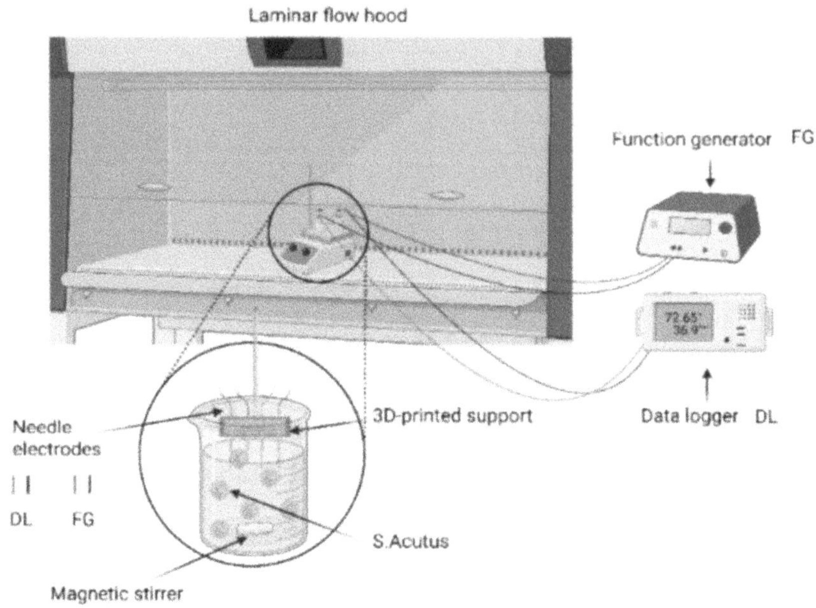


**Fig. 2.** Experimental setup for long-term monitoring of the electrical activity of *S. acutus.* All recordings were carried out within a biological safety cabinet to ensure a sterile and controlled environment.

The culture medium without *S. acutus* was used as a reference, serving as a blank control for comparative analysis. A custom 3D-printed support was designed to hold the electrodes within the suspension and to integrate additional electrodes for electrical stimulation, connected to an arbitrary waveform generator (Rigol DG1022Z). The analyzed data were obtained from experiments conducted on cultures of *S. acutus* lasting approximately 3 h, containing about 11,000 data points, sampled at 1 Hz.

## 3.2 Computational Analysis Tools

For numerical analysis, we employed different techniques: i) Frequency Domain Analysis tools such as Fast Fourier Transform (FFT) and periodogram [7] to identify characteristic frequencies and periodicities present in the voltage signals, which could be related to biological cycles or responses to environmental conditions (e.g., implicit light/dark cycles or other metabolic oscillations), ii) Spectrograms [8, 9] to obtain a time-frequency representation of the signals, allowing us to observe how the spectral content changes over time and to locate transient or non-stationary events, and iii) Autoencoders based on Recurrent Neural Networks (RNNs), particularly those based on Long Short-Term Memory (LSTM) cells.

As regards FFT, this is an efficient algorithm for computing the Discrete Fourier Transform (DFT). The FFT decomposes a discrete signal into its constituent frequency components, allowing for the analysis of the spectral content of complex signals in various fields such as audio processing, communications, and biomedical signal analysis

[10]. Linked to the FFT, the periodogram is a standard method to estimate the power spectral density (PSD) of a signal. It can be computed by calculating the squared magnitude of the FFT of the signal: this provides an estimate of how power is distributed across the various frequencies composing a signal. Despite its simplicity and its computational efficiency, the periodogram can produce high-variance estimates; hence, modified versions such as the Welch method are often employed to improve its stability [11].

A Spectrogram extends the periodogram concept to non-stationary signals by computing the FFT over short, overlapping time windows, creating a time–frequency representation. This feature enables the visualization of how the frequency content evolves over time. The spectrogram is widely used in machine learning applications, particularly as an input representation for audio and speech recognition systems [12].

On the other hand, RNNs are a class of artificial neural networks aimed at processing sequential data by maintaining an internal state that captures temporal dependencies. However, standard RNNs suffer from vanishing and exploding gradient problems when modeling long sequences; thus, LSTM networks were introduced [13]. This type of neural network incorporates gating mechanisms, i.e., input, output, and forget gates, regulating information flow, and making the model capable of learning long-term dependencies effectively. These architectures have become foundational in time-series forecasting, such as speech recognition, and natural language processing.

Finally, an LSTM-based autoencoder combines the temporal modeling capability of LSTMs with the structure of an autoencoder, a neural network designed to learn compressed representations of data. In the considered configuration, the LSTM encoder processes sequential inputs and compresses them into a latent state or representation, whereas the LSTM decoder tries to reconstruct the original sequence from this latent code. When trained on normal or known patterns, the model learns to reconstruct them accurately. During inference, unknown or anomalous patterns, i.e., those that deviate from the learned dynamics, lead to higher reconstruction errors, enabling effective unsupervised anomaly or unknown pattern detection in time-series and signal data [14].

All the aforementioned models and tools have been implemented by using Python version 3.10 and the NumPy, SciPy, as well as TensorFlow libraries in a Windows 11 environment.

## 4 Experimental Results

In this section, we report the results obtained applying the methodologies described in Sect. 3.2 to the signals coming from both the algae and the medium in the two considered scenarios: without or with an electrical stimulus.

### 4.1 Time and Frequency Analysis

In this subsection, we present the results regarding the time and frequency analysis using FFT and periodograms.

Figure 3 compares the voltage time series of *S. acutus* (blue trace) and the culture medium (red trace) in the absence (left panel, *Not Stimulated*) and presence (right panel, *Stimulated*) of electrical excitation. Under resting conditions (Fig. 3 on the left), *S.*

*acutus* shows a weak oscillatory behavior, whereas the reference medium remains almost flat, displaying no significant voltage variations over time. When electrical stimuli are applied (Fig. 3 on the right), the response of *S. acutus* becomes markedly enhanced: the oscillations increase in amplitude and appear superimposed on a transient baseline shift, indicating a clear rise in the electrical activity of the organism. In contrast, the medium continues to exhibit a nearly constant potential throughout the entire measurement. This distinct behavior suggests that *S. acutus* possesses an intrinsic electrogenic activity that can be triggered or amplified by external electrical fields, possibly reflecting a coupling between the applied stimulus and ionic or membrane-level processes within the alga, evoking a primitive form of electrical communication emerging from its cellular structure.

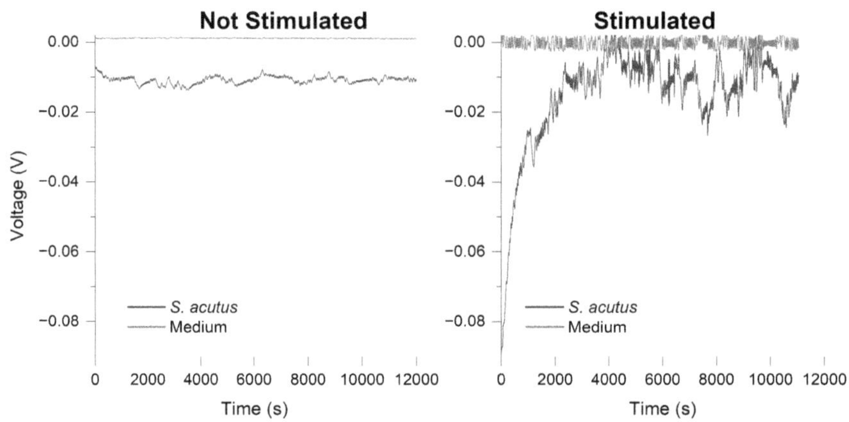


**Fig. 3.** The voltage time series of *S. acutus* (blue line) and the culture medium (red line) recorded in the absence (left panel) and presence (right panel) of electrical stimulation. Under stimulation, *S. acutus* exhibits a marked increase in oscillatory activity, indicating an enhanced electrogenic response, while the medium remains almost constant.

FFTs and periodograms (Figs. 4 and 5, respectively) identified persistent low-frequency signals, likely related to metabolic cycles or environmental conditions. This behavior becomes neater and more evident when an electrical stimulus is applied, with an increase of the amplitude of the order of about ten times for the FFT and a hundred times for the periodogram, which, moreover, exhibits a clear frequency peak in the stimulated conditions.

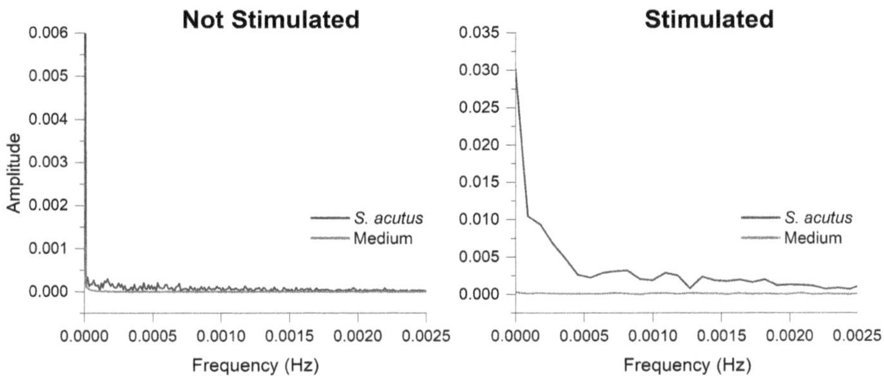


**Fig. 4.** Fast Fourier Transforms of both medium (red) and *S. acutus* (blue) with (right) or without (left) electrical stimuli.

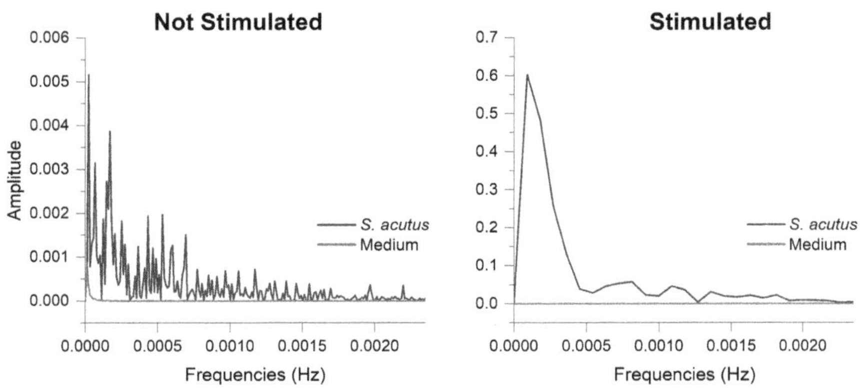


**Fig. 5.** Periodograms of both medium (red) and *S. acutus* (blue) with (right) or without (left) electrical stimuli.

## 4.2 Spectrograms

In this subsection, we briefly describe the results regarding spectrograms. Particularly, Figs. 6 and 7 show the spectrograms of the medium and *S. acutus* without or with the application of an electrical stimulus, respectively.

As one can see, Fig. 6 demonstrates that *S. acutus* spontaneously exhibits a series of low frequencies, between 0 and 0.05 Hz, throughout the full time recording of about 3 h, with a relevant amplitude, in particular, in some time intervals: very first minutes, between minutes 50 and 60, minute 72, minute 100, etc. This behavior is not present or present at a very very low degree, compared with the algal behavior, in the medium alone (bottom part of Fig. 6).

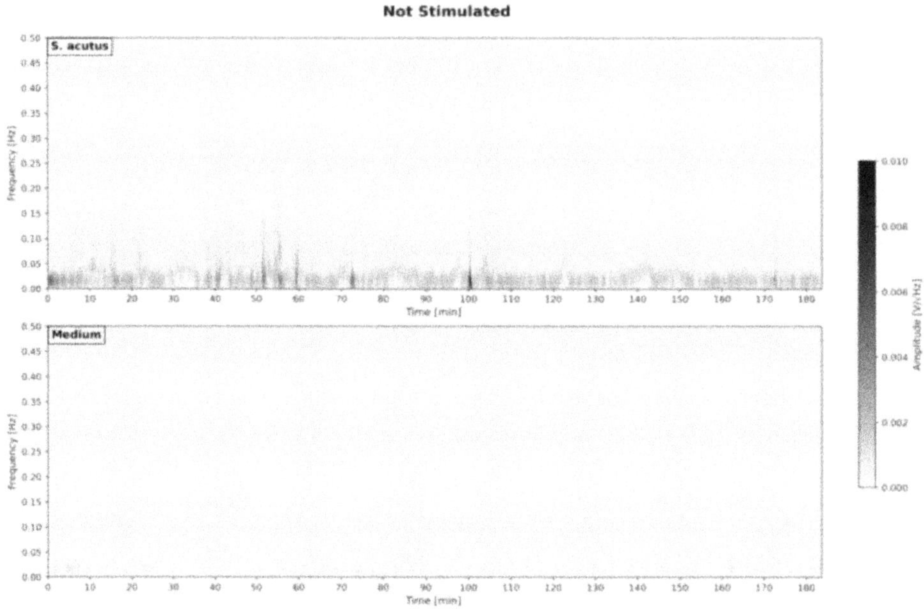


**Fig. 6.** Spectrograms of both medium (bottom) and *S. acutus* (top) regarding the spontaneous electrical activity.

On the other hand, in Fig. 7, we show the spectrograms of both the medium (bottom part) and *S. acutus* (top part) when the considered electrical stimulus (0.1 Hz, 10mV) is applied. As one can clearly see, the frequency patterns shown in Fig. 6 are exacerbated both in the medium and in the algal activity. However, while in the medium the frequencies are confined within 0.1 Hz and with a much smaller amplitude, in the algal spectrogram various spectral patterns appear around 0.1 Hz at minutes 80, 85, 90, 105, for example. Moreover, at the same timestamps, further frequencies acquire a relevant amplitude such as 0.3 Hz and 0.5 Hz. Another interval between about 130 and 150 min, not present in the medium spectrogram, appears clearly in the algal spectrogram, always at a frequency equal to 0.1 Hz. Finally, in the 160–170 min interval, both the medium and the algal spectrogram show some highlight on the 0.1-Hz frequency band, but the one of the algae is evidently stronger, confirming once more the better reaction of *S. acutus* to the provided stimulus with respect to the medium itself.

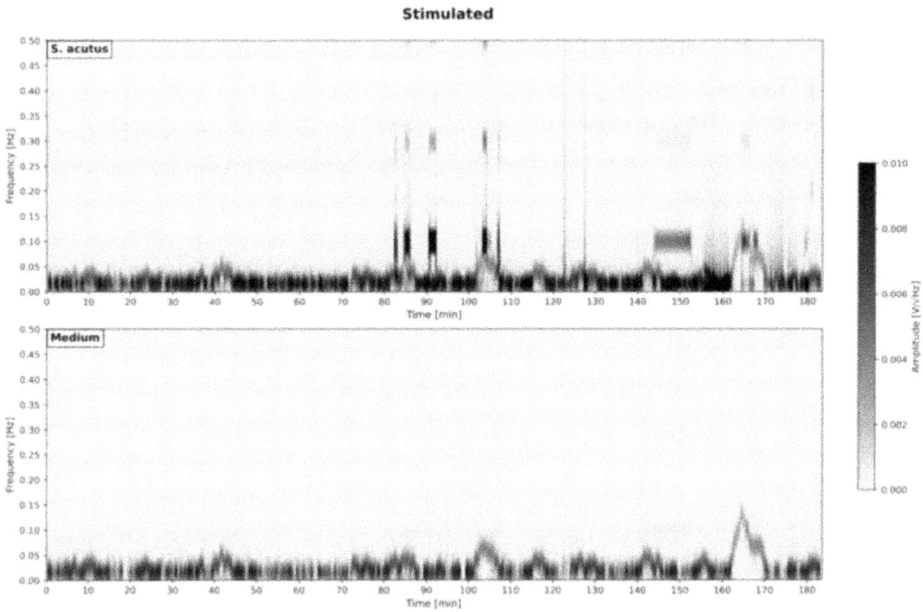


**Fig. 7.** Spectrograms of both medium (bottom) and *S. acutus* (top) when applying an electrical stimulus.

### 4.3 LSTM-Based Autoencoder Results

In this subsection, we briefly describe the results regarding the intervals found through the application of the LSTM-based Autoencoder. First of all, we describe the main hyperparameters of the considered neural network (see Table 1) and then we show the obtained outcomes, for the stimulated scenario, in Table 2.

As regards the hyperparameters, we considered the following values obtained through grid search optimization: a window size equal to 500 in order to capture longer patterns despite the slower training time; a stride equal to 10 for smoother processing; a balanced encoding dimension, i.e., the size of the latent representation, equal to 20, between stronger compression and the preservation of more information; two layers of LSTM hidden units with standard values; a number of epochs equal to 300 but with early stopping enabled; a batch size equal to 64 for faster training; and an intermediate learning rate equal to 0.001 for the Adam optimizer. Moreover, we considered a 20% validation split rate and a threshold percentile, i.e., the percentile of anomaly scores to use as a threshold, equal to 75%, rather small to increase the sensitivity of the method of detecting anomalous patterns.

**Table 1.** Hyperparameters of the considered LSTM-based Autoencoder.

| Hyperparameter | Value |
|---|---|
| Window Size | 500 |
| Stride | 10 |
| Encoding dim | 20 |
| Learning rate | 0.001 |
| Batch size | 64 |
| Dropout rate | 0.2 |
| LSTM units | [64, 32] |
| Epochs | 300 |

As concerns the detected temporal intervals, they are shown in Table 2 as for the stimulated scenario. In the table, we report the starting minute in the about 3-h recording, the duration of the oscillatory pattern, and the average frequency in the pattern itself. As one can see, the reported intervals match the outcomes obtained through the spectrograms previously shown.

**Table 2.** Some relevant intervals detected through the LSTM-based Autoencoder in the stimulated scenario.

| Start time (min) | Duration (min) | Mean frequency (Hz) |
|---|---|---|
| 91.5 | 3.16 | 0.10 |
| 104.3 | 1.42 | 0.11 |
| 144.2 | 7.95 | 0.09 |

## 5 Conclusions and Future Directions

In this paper, we have analyzed the spontaneous and stimulated electrical activity generated by *Scenedesmus acutus* cultures by using different computational and analysis tools. We have highlighted temporal patterns and characteristic frequencies in its spontaneous electrical activity, which can be further amplified with the application of proper electrical stimuli. These spectral patterns suggest that microalgae could serve as biological generators of electrical signals or components of future bio-hybrid systems, and they could potentially be used for the realization of bio-electric circuits. However, further investigations are certainly needed to identify the environmental conditions that most influence their electrical activity to integrate algae into non-conventional sensing or computing architectures, by exploiting their natural dynamics in bio-inspired systems that can, in turn, be used, for example, for environmental sensing, bio-computing, sustainable sensing, etc.

Future research directions may regard the study and comparison of the electrical activity of different types of algae, not limiting to *S. acutus*, the application of different computational methods to detect the presence of patterns in the electrical signals of the algae, as well as the assessment of the application of different types of stimuli, i.e., mechanical, sound-based, light-based, etc. to the solution containing the considered algae.

**Acknowledgments.** This work has been supported by the National Recovery and Resilience Plan (NRRP), Mission 04 Component 2 Investment 1.5 – NextGenerationEU, Call for tender n. 3277 dated December 30, 2021, Award Number: 0001052 dated June 23, 2022. Prof. Riccardo Pecori acknowledges INdAM GNCS research group.

**Disclosure of Interests.** The authors have no competing interests to declare that are relevant to the content of this article.

## References

1. Monk, T., et al.: Electrical signaling beyond neurons. Neural Comput. **36**(10), 1939–2029 (2024)
2. Phillips, N., Mayne, R., Adamatzky, A.: Chlorella sensors in liquid marbles and droplets. Sens. Bio-Sens. Res. **36**, 100491 (2022). https://doi.org/10.1016/j.sbsr.2022.100491
3. Antonacci, A., Scognamiglio, V.: Biotechnological advances in the design of algae-based biosensors. Trends Biotechnol. **38**(3), 334–347 (2020). https://doi.org/10.1016/j.tibtech.2019.10.005
4. Brayner, R., Couté, A., Livage, J., Perrette, C., Sicard, C.: Micro-algal biosensors. Anal. Bioanal. Chem. **401**(2), 581–597 (2011). https://doi.org/10.1007/s00216-011-5107-z
5. Borowitzka, M.A., Beardall, J., Raven, J.A. (eds.): The Physiology of Microalgae. Springer, Heidelberg (2016). https://doi.org/10.1007/978-3-319-24945-2
6. Marieschi, M., Gorbi, G., Zanni, C., Sardella, A., Torelli, A.: Increase of chromium tolerance in Scenedesmus acutus after sulfur starvation: chromium uptake and compartmentalization in two strains with different sensitivities to Cr(VI). Aquat. Toxicol. **167**(2015), 124–133 (2015). https://doi.org/10.1016/j.aquatox.2015.08.001
7. Bruns, A.: Fourier-, Hilbert- and wavelet-based signal analysis: are they really different approaches? J. Neurosci. Methods **137**(2), 321–332 (2004). https://doi.org/10.1016/j.jneumeth.2004.03.002
8. Mallat, S.: A Wavelet Tour of Signal Processing: The Sparse Way, 3rd edn. Academic Press (2008). ISBN: 9780123743701
9. Cohen, M.X.: Analyzing Neural Time Series Data: Theory and Practice. MIT Press (2014). https://doi.org/10.7551/mitpress/9609.001.0001
10. Cooley, J.W., Tukey, J.W.: An algorithm for the machine calculation of complex Fourier series. Math. Comput. **19**(90), 297–301 (1965)
11. Welch, P.D.: The use of fast Fourier transform for the estimation of power spectra: a method based on time averaging over short, modified periodograms. IEEE Trans. Audio Electroacoust. **15**(2), 70–73 (1967)
12. Oppenheim, A.V., Schafer, R.W.: Discrete-Time Signal Processing. Pearson (2009)
13. Hochreiter, S., Schmidhuber, J.: Long short-term memory. Neural Comput. **9**(8), 1735–1780 (1997)
14. Malhotra, P., Vig, L., Shroff, G., Agarwal, P.: Long short term memory networks for anomaly detection in time series. In: ESANN 2015 Proceedings, pp. 89–94 (2016)

# Going Deep in the Origin and Diversity of Life

# Collective Autocatalysis in a Modified Binary Polymer Model

Samuele Carpi[1], Roberto Serra[1,2,3], and Marco Villani[1,2](✉)

[1] Department of Physics, Informatics and Mathematics, University of Modena and Reggio Emilia, Modena, Italy
marco.villani@unimore.it
[2] European Centre for Living Technology, Venice, Italy
[3] Institute of Advanced Studies, University of Amsterdam, 1012 WX Amsterdam, The Netherlands

**Abstract.** This study explores the impact of incorporating structure-dependent catalysis into a polymer chemistry model initially proposed by Kauffman to investigate autocatalytic systems and the origin of life. The research reveals significant changes in chemical organization and autocatalytic set formation: in the modified model (C-chemistries), a few catalysts can drive many reactions, unlike the chemistries created by the Kauffman model. C-chemistries are either small with low connectivity or large and highly connected; the latter always contain large reflexively autocatalytic and food-generated sets (RAFs), while small C-chemistries can also have RAFs, though less frequently. The model shows that increasing the cleavage fraction reduces the likelihood of RAF formation. Importantly, the study suggests that large RAFs face synchronization challenges, a phenomenon relevant to understanding life's emergence and warranting further investigation.

**Keywords:** Autocatalytic systems · protocells · origin of life · structure-dependent catalysis · BPM

## 1 Introduction and Goals

A striking feature of biology is the widespread presence of long polymer chains, like proteins and nucleic acids, which are necessary for survival, growth and reproduction. In present-day organisms the syntheses and transformations of these polymers are strongly regulated by sophisticated control mechanisms [1], which were unlikely to have been at work in primitive protocells that are supposed to have existed at the dawn of life [2]. Hypotheses about the way in which polymers were synthesized and transformed are therefore key topics in studies on the origin of life (OOL) [3–7].

Today's long chains represent the outcome of long evolution processes, where in the beginning there were only monomers or very short oligomers. Most papers on the synthesis of long polymers rely on definite choices of specific chemicals (e.g. oligopeptides or oligonucleotides) as the starting species, however there are still large uncertainties


M. Bianchini et al. (Eds.): WIVACE 2025, CCIS 3001, pp. 89–105, 2027.
https://doi.org/10.1007/978-3-032-33185-4_7

about the chemical identity of the first long biological polymers. It is therefore interesting to consider also the behavior of sets of randomly interacting molecules: as it is often the case, this approach may be helpful to clarify generic properties, which are held under a number of different specific hypotheses. "Generic models" of this kind are widespread in physics, and they have proved fruitful also in biology [8, 9] and in OOL [4, 10, 11].

A model introduced about 40 years ago by Stuart Kauffman [3], often called the binary polymer model (BPM), has raised much attention in studies on OOL. The abstract "polymers" of the BPM are binary strings, since it is supposed that there are two kinds of building blocks, which can represent two different monomers or perhaps two different sets of monomers, like hydrophilic or hydrophobic ones. For simplicity, we will always refer to the building blocks as monomers. There are two operators which can modify the polymers: cleavage, where one string is cut, giving rise to two new shorter polymers, and condensation, where the ends of two strings are joined to form a longer one. It is assumed that reactions can take place only if they are catalyzed. Besides being a reactant, each polymer can be a catalyst of some randomly chosen reactions.

The key questions concern how the reaction network grows in time, from an initial small set of different molecular types to a larger set of chemical species. There are basically two different kinds of studies, which can be called for simplicity "topological" and "truly dynamical", both starting from an initial set of polymer types. In the former case, one studies how the reaction network grows as new species are produced, supposing that all the reactions which are possible, given a certain set of reactants and catalysts, do actually take place (thus ignoring the constraints of chemical kinetics) [3]. More precisely: at each iteration, a catalyst is chosen at random, and all the reactions that it catalyzes do take place. In this way a new set of polymer types may come into existence, and at the next step another catalyst is chosen etc.

In the truly dynamical approach, only those reactions take place whose reactants and catalysts are present (there is a numerical variable associated to each chemical species, which measures its density). The strings of the BPM are treated as if they were true chemicals, and their densities change according to kinetic laws. The rate of a reaction is determined by the usual law of mass action, so it is proportional to the frequency of encounters of the molecules which partake in that reaction, i.e. reactants and catalyst. The topological type of analysis allows one to determine which species can possibly be found, and which reactions can take place, i.e. the "artificial chemistry" which covers all the possible stories which can be observed in that "world". The truly dynamical analysis also takes into account the densities of the various chemical types, so specific stories can be observed [4, 11–13].

In the original, "classical" BPM - which will be called K-BPM, its artificial chemistries will be called K-chemistries, etc. - each polymer has a certain probability of catalyzing a reaction, which does not depend upon its structure (i.e. the sequence of its monomers) but is chosen at random. If some species are assumed to be freely available in the environment (the "food") then it is possible to study the growth of the artificial chemistry. It is also important to study the formation of RAFs (Reflexive Autocatalytic Food-generated sets) of chemicals and reactions, which can be self-sustaining. RAFs provide a convenient formalization of the notion of collectively autocatalytic sets (see [14, 15] for a precise definition), which takes also into account the need for some freely available food to assure self-replication. The "topological" approach is well suited to study this type of problems. By assigning at random catalysts to reactions, i.e. defining an artificial chemistry, one can study under which conditions RAFs can actually appear (e.g., as a function of the properties of the initial set of polymers).

While this approach ignores how many specimens of a given type are actually there, the second approach tries to determine the dynamical behavior of a system, starting from a certain initial set of quantities of molecules, not just from knowledge of their types. Both approaches provide useful information. In this paper we will introduce a new version of the BPM, and we will extensively analyze its properties with the "topological" approach. We will also mention some preliminary results of the "truly dynamical" kind, deferring a more complete treatment to further works.

While the simplifications of generic models can always be questioned, in the case of the k-BPM there is one feature which seems to require much care, i.e. the random association of catalysts to reactions, irrespective of their structure. This is at odds with actual chemistry, therefore in this paper we explore the behavior of a modified BPM (which will be called C-BPM, its artificial chemistries will be called C-chemistries, etc.), where the catalytic properties are associated to the sequence of monomers. For example, using complementary base pairing, we may suppose that the cleavage of a polymer containing the sequence ABAAB is catalyzed by a polymer whose "active site" hosts the sequence BABBA.

Of course, present-day biological polymers are made of more than two different types of monomers (for example, there are four different types of monomers in DNA molecules). However, we wish to keep the model as simple as possible, therefore we will limit it to binary polymers, since we deem it necessary that different polymers of the same length can exist (and two is the smallest number of types of building blocks that make it possible). Generalizations to other cases, with a larger number of different monomers, could be straightforward, but it would require defining the laws for complementary pairing, which are straightforward in the binary case. Let us also observe that the most interesting properties of the model are likely to be unaffected by the choice of a different matching rule (for example, coupling identical rather than complementary monomers – of course, assuming that an active site cannot correspond to itself).

A similar model has been introduced by [16]. Here we provide a more thorough analysis of its behavior. We analyze under which conditions random interacting polymers can give rise to RAFs and how these RAFs compare with those of the standard K-BPM. We will also provide preliminary results about the problem of identifying under which conditions these RAFs can actually survive and replicate in sustainable populations of random protocells, which requires a "truly dynamical" approach. For this purpose, rather than the usual flow reactors we will consider the model described in [11], which is well suited to deal with growing and dividing protocells.

## 2 The Modified BPM

In the BPM, the entities are binary strings of length $L$, whose elements (monomers) are taken from a two-letter alphabet, e.g. {A,B}. The sequence matters, so AB is different from BA: there are therefore $2^L$ different types of polymers which are $L$ monomers long. Each sequence is a different chemical type (or species). A cleavage reaction cuts a polymer into two shorter pieces, provided that the catalyst for this reaction exists. If the polymer $P$ which undergoes cleavage is composed by two parts $P_1P_2$, and if $C$ is the catalyst for that reaction, then

$$P_1P_2 + C \rightarrow P_1 + P_2 + C$$

A condensation reaction joins two polymers $Q$ and $R$ in a longer one $QR$:

$$Q + R + C \rightarrow QR + C$$

Note that reactions in the BPM are unidirectional. The "artificial chemistry" is dynamically generated by applying, one at each iteration step, to all the possible reactions (i.e. the cleavages and condensations for which a catalyst is present) to a set of existing polymers. In this way a larger set of chemical species is generated, and new reactions become possible. By applying all the possible reactions to this larger set, a further set is obtained, and so on.

Condensation can give rise to longer and longer molecules: since the number of possible reactions is an exponential function of the number of different species, it is computationally necessary to limit their maximum length. This can also be "chemically" justified by the fact that molecular weight is an increasing function of length, and very long molecules may precipitate and become no longer available for further reactions (which are implicitly assumed to take place in a fluid). All this translates into a limit on the maximum length $L_{max}$ of a molecule, so reactions leading to larger molecules are prohibited. The length of the largest "legal" polymers is therefore $2^{Lmax}$ and the maximum number of legal polymers is $2^{Lmax+1}$-2. In some cases, the parameters and initial conditions are such that very long species are not generated, thus making the limitation irrelevant. The fact that, depending upon initial conditions and model parameters, there may be intrinsic limitations to the number of species and reactions is an interesting feature that deserves consideration (see Sect. 3). We will refer to the final number of species as the "size" of the artificial chemistry associated to a particular realization of the model.

The classical K-BPM and the modified C-BPM differ by the way in which catalysts are associated to the reactions which they catalyze: in the former case, this is done at random, usually with uniform probability (but some variants have also been considered where the probability depends on the length of the molecule [16]). In C-BPM model, it is assumed that each catalyst can host one "active site", which can match a portion of one or two polymers, catalyzing a cleavage or a condensation reaction.

Some specific choices are needed to precisely define the model; in this paper we will assume that, whenever a new species is generated, it is either labelled as a catalyst (for cleavage or condensation) or not. If it is a catalyst for cleavage, a substring of length $\Lambda \leq L$ will be identified as its "active site" and all the existing polymers will be scanned searching for the complementary substrings (if any). A cleavage reaction will then be considered at each matching position, possibly generating some new species. If the newborn polymer is labelled as a catalyst for condensation, all the pairs of polymers will be considered, looking for those cases where concatenation of the end part of a molecule $M_1$ and of the initial part of another $M_2$ matches the active site, in this case a condensation takes place, possibly generating new species $M_1M_2$ (see Fig. 1). In this approach, only polymers at least $\Lambda$ long can be catalysts.

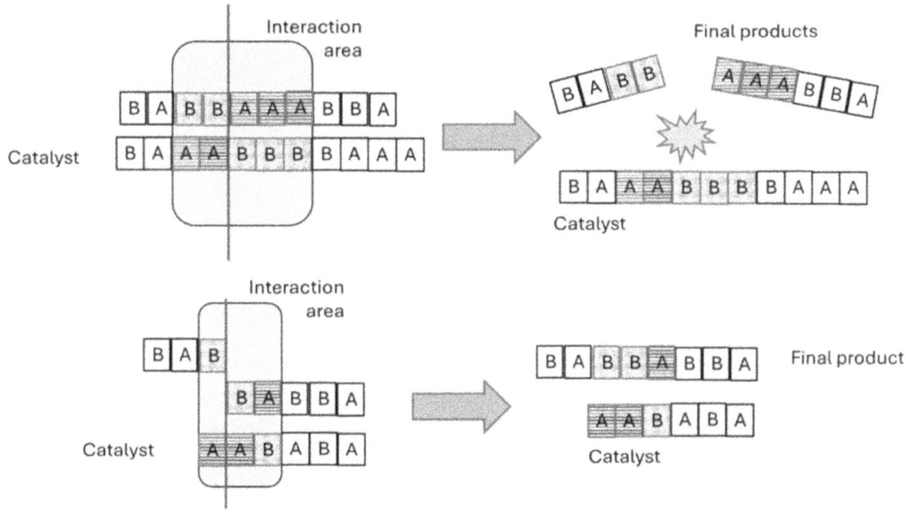


**Fig. 1.** Structure-based examples of cleavage and condensation. Top: cleavage. An active site (colored monomers) and a cut point are identified in a catalyst. A chemical species becomes a reagent if a substring complementary to the active site of the catalyst is identified within it: in this case, the chemical species is broken at the cut point. Bottom: condensation. An active site (colored monomers) and a suture point are identified in a catalyst. A chemical species becomes the first reagent if its terminal part is complementary to the first part of the active site, while a chemical species becomes the second reagent if its initial part is complementary to the second part of the active site.

In order to start the iterative procedure for generating artificial chemistries, the initial conditions must be defined. In various papers on the BPM, the initial condition is composed of all the monomers and of all the short oligomers, whose length does not exceed a threshold $h_1$ (see Table 1). In this paper, the initial condition is usually composed of all monomers, all the short oligomers, whose length does not exceed a threshold $h_1$ and of a fraction of larger polymers, whose length is larger than $h_1$ and smaller than a second threshold $h_2$. The values of the parameters are given in Table 1. This initial set of chemical species is sometimes referred to as the "firing disk". The idea is that the environment provides some small pieces which can be joined to produce larger types; moreover, cleavage assures that large polymers can be broken into pieces (a schematization in accordance with hypotheses regarding primitive prebiotic environments [17], such as the "warm little pound" by Darwin [18] or the "primordial soup" of Oparin [19] and Haldane [20]). Figure 2 shows the first steps in creating K-chemistries and C-chemistries.

It is assumed that a small fixed number of randomly chosen initial polymers are catalysts for cleavages and that the same number of polymers are catalysts for condensations (see Table 1). A catalyst is chosen at random among the existing ones, and the

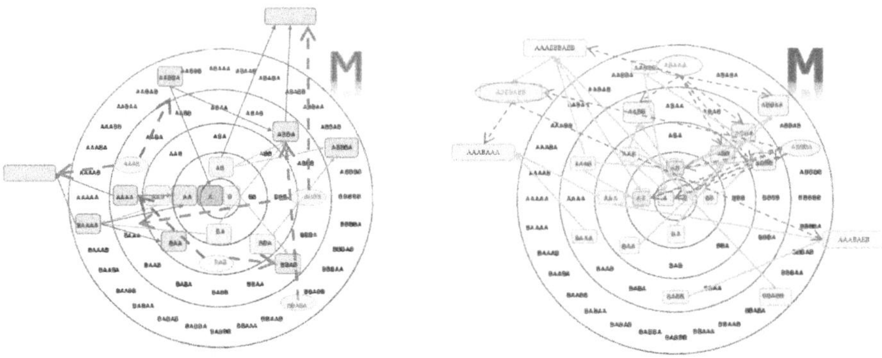


(a) (b)

**Fig. 2.** The first steps in constructing K-chemistries (a) and C-chemistries (b). Catalysis links (who catalyzes what) are indicated by red dashed arrows, while production links (who is the building block for what) are indicated by solid orange arrows. In the K-chemistry on the left, each reaction typically has a different catalyst with a structure unrelated to that of the reactants. If a catalyst - rarely, and only by chance - catalyzes more than one reaction, the reactions in question have no structural links between them. Condensation reactions can create products whose length exceeds that of the chemical species initially present (the maximum diameter of the "firing disk"). The final situation is such that a large number of reactions requires a large number of catalysts. In the C-chemistry on the right, each catalyst typically catalyzes more than one reaction, each one activated by the same active site. The final situation is such that a few catalysts can catalyze a large number of reactions.

initial population is searched, looking for complementary matches. When one is found, the corresponding reaction takes place, possibly generating new species. Whenever a new polymer with $L \geq \Lambda$ is generated, it is randomly chosen to be a catalyst with a small fixed probability $p_{cat}$. As mentioned above, a catalyst can catalyze either a cleavage or a condensation; the fraction of catalysts for cleavage $p_{cl}$ varies in different experiments (the fraction of catalysts for condensation being of course 1- $p_{cl}$). Note that the probability that a new polymer is a catalyst for cleavage is equal to $p_{cat}p_{cl}$. The procedure is iterated until no new "legal" species can be formed, or some termination condition is met.

Finally, let's add a note regarding the "specificity" of reactions in models and in scenarios regarding the origin of life. It is thought that in the first pre-biotic systems the specificity of catalysts was not very high [21, 22]: often the same catalyst was able to accelerate entire classes of reactions - even if only slightly - since its action was of a generic type. In current systems, after billions of years of evolution, biological catalysts (enzymes) are extremely powerful and specific: they are able to accelerate the rate of a single reaction, even millions of times compared to the uncatalyzed reaction [23, 24]. This latter situation is captured by the original Kauffman model, in which each catalyst typically catalyzes only one reaction, or a few selected reactions. In the C-BPM model the action is carried out through the active site: any reaction whose reagents are coherent with that site is possible. Obviously, the larger the active site, the greater the degree of specificity; but even with the parameters used in Table 1, where the size of the active site is not a small fraction of the total length of the molecule, each catalyst catalyzes a large number of reactions (as we will discuss extensively later in the paper).

**Table 1.** The table summarizes the main parameters of the model, and those concerning the initial setting. The symbols used, a brief description of the parameter, and some typical values used during the simulations are reported.

| Model | Symbol | Description | Value during experiments |
|---|---|---|---|
| Initial conditions | $h_1$ | maximum length of species surely present in the firing disk | 2 |
| | $h_2$ | maximum length of species present in the firing disk | 4 |
| | $NCL_{ini}$ | number of cleavage catalysts present in the firing disk | 2 |
| | $NCD_{ini}$ | number of condensation catalysts present in the firing disk | 2 |
| C-chemistry | $L_{max}$ | maximum length of species belonging to chemistry | 10 |
| | $\Lambda$ | length of the active site (may be different in different species) | Uniformly distributed between 3 and 4 |
| | $p_{cat}$ | probability that a new chemical species is a catalyst | Typically, 0.05 |
| | $p_{cl}$ | the fraction of catalysts for cleavage | Within the range [0.1,0.9] |

## 3 Simulation Results

Let us recall that the size of an artificial chemistry is the number of different types which are found at the end of the simulation, and that the set of "legal" types is composed by all the binary polymers not longer than $L_{max}$. A striking result is that in C-BPM two very different behaviors are observed: the size of a chemistry is either small (much smaller than the number of legal types, and only slightly larger than the size of the initial set of species) or quite large, when the number of different species approaches that of all the legal types. In these latter cases one also finds a very high connectivity (defined here as the ratio between the number of reactions and the number of species). Intermediate-size chemistries are practically absent. If the probability that the new polymer is a catalyst is low (with current parameters, less than 0.03), then we find almost only small chemistries; if we use a slightly larger catalysis probability (with current parameters, greater than 0.1), we find almost only large chemistries (Fig. 3).

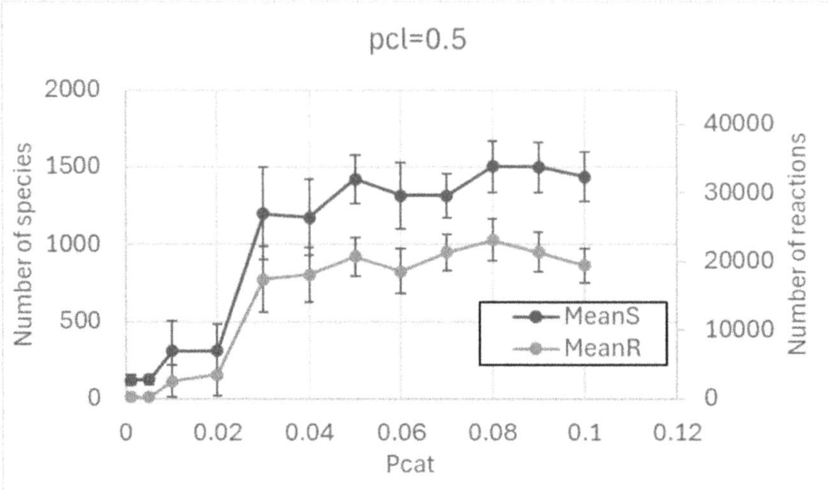


**Fig. 3.** The mean number of chemical species and the mean number of reactions in ensembles of 100 networks, maintaining equal proportions of cleavages and condensations. There is a region ($p_{cat} < 0.01$) where almost exclusively small-sized chemistries are found, and a region ($p_{cat} > 0.1$) where almost exclusively large-sized chemistries are found. In the interval between these two values it is possible to find both large and small chemistries (intermediate-sized chemistries being almost absent).

There is a narrow range of intermediate values of $p_{cat}$ where both behaviors can be observed. Interestingly, even in these cases one usually observes either large chemistries or small ones, while intermediate values have very rarely observed. Moreover, the choice of the fraction of catalysts which catalyze one type of reaction heavily affects the observed behavior, as shown in Fig. 4, where one can see that, the higher the fraction of cleavages, the lower the probability of obtaining large-size chemistries. This can be understood by observing that condensations are the operators which are able to generate new long species: if they are absent or very rare, only relatively small polymers can ever form in significant numbers.

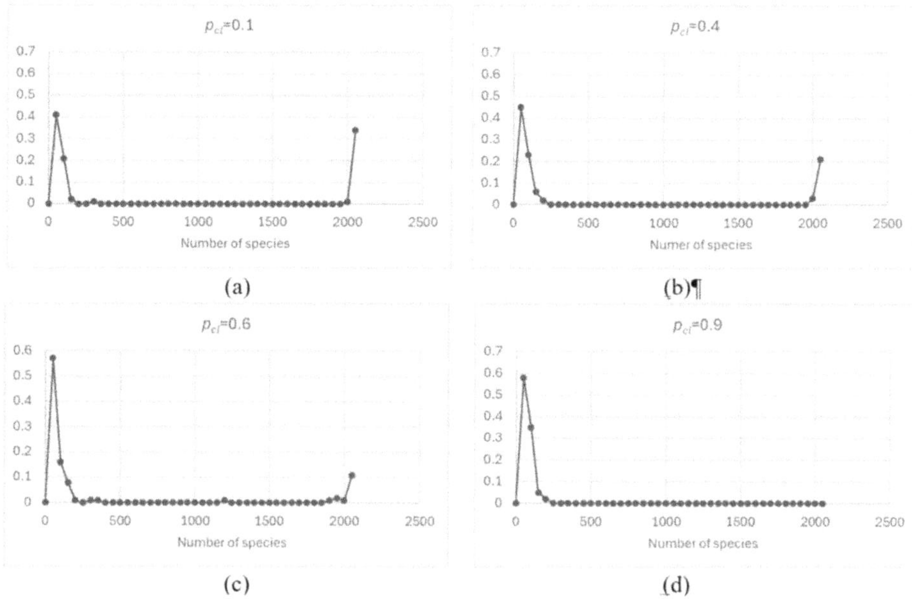


**Fig. 4.** Chemistry size distributions (number of chemical species) as the cleavage fraction varies (cleavage at 0.1, 0.4, 0.6 and 0.9, respectively). It can be noted that typically there are no chemistries of intermediate size between the "small" and "large" peaks. Large chemistries are more likely when the condensation fraction is prevalent.

The sharp bimodality allows one to separately analyze the behavior of large and small chemistries. While keeping it constant the probability $p_{cat}$ that a newborn polymer is catalyst, it is interesting to consider the effects of changing the probability $p_{cl}$ that a cleavage is chosen (the probability that a new polymer is a catalyst for cleavage is equal to $p_{cat}p_{cl}$). As shown in Fig. 5b, the size of small chemistries remains almost constant; this happens in a wide range also to large chemistries, whose numbers however quickly vanish as $p_{cl}$ approaches 0.9.

The number of possible reactions in a BPM is an exponentially growing function of the number of species (unless the boundaries are reached) therefore the typical connectivity of large chemistries is much higher than that of the smaller chemistries (Fig. 6a). Furthermore, in the simulations which have been performed, all large chemistries contain a RAF, whose size is equal to, or slightly smaller than that of the chemistry itself (Fig. 6d).

On the contrary, only a fraction of the small chemistries contains a RAF (Fig. 6b), whose size is always smaller than the chemistry itself. Consequently, RAFs contained in large chemistries are themselves large and have high connectivity, while RAFs contained in small chemistries are typically smaller than the containing chemistries, and have similar (low) connectivity. The number of RAFs in large chemistries decreases as $p_{cl}$ increases (Fig. 6c) as does the number of large chemistries (Fig. 5a).

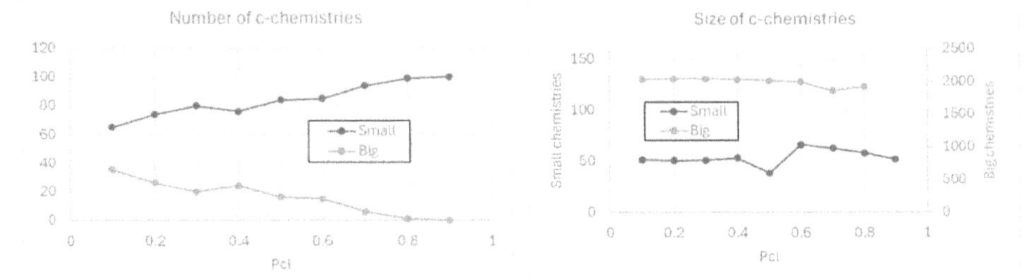


(a) (b)

**Fig. 5.** (a) Number of large and small chemistries, as the cleavage probability varies. (b) Average size of the two different groups of chemistries as the cleavage probability varies. The large chemistries scale is the one on the right, the maximum possible value being $[2^{11}–2] = 2046$ ($L_{max} = 10$) chemical species: the series stops at $p_{cl} = 0.9$, a situation in which no large chemistries are detected (100 launches for each point).

Certainly, the organization of the C-chemistries is very different from that of K-chemistries, produced by the classical Kauffman algorithm. In order to provide meaningful comparisons, in this work we compare chemistries that have the same connectivity. Note however that the typical connectivity of small C-chemistries is really small (about one reaction per species, see Fig. 6a). Since it is known from previous studies that K-chemistries with such a low connectivity typically do not host any RAF [11], in the following we focus on large K-chemistries only.

By comparing a C-chemistry to a K-chemistry with the same connectivity (let us recall that it is defined as the ratio between the total number of reactions and the number of species), when cleavages are as likely as condensations ($p_{cl} = 1/2$) one typically observes that:

a) A C-chemistry has a high number of reactions, despite a low number of catalysts (typically, about 100). To achieve connectivity levels comparable to those observed in such large C chemistries, all the species in a K-chemistry must catalyze several reactions. Consequently in any C-chemistry the number of reactions catalyzed by each catalyst is significantly higher than the number of reactions catalyzed by each catalyst in a K-chemistry
b) The distribution of the number of reactions catalyzed by each catalyst in a K-chemistry takes the form of a bell curve, while in a C-chemistry there it takes an irregular shape (Fig. 7)
c) In C-chemistry, the average number of reactions catalyzed by each catalyst (400 reactions) is much higher than the average number of species catalyzed (210 species) – the opposite situation to that in K-chemistry, where the average number of reactions catalyzed by each catalyst (19.6 reactions) is lower than the average number of species catalyzed (26.8 species)

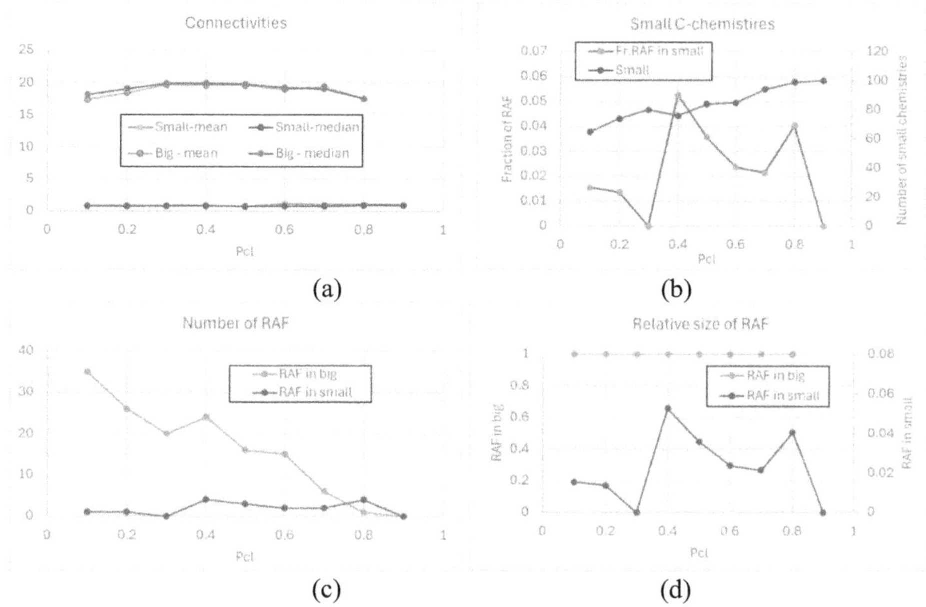


**Fig. 6.** (a) Average and median connectivity of large-scale and small-scale chemistries. (b) Number of small-sized chemistries (out of 100 runs, right-hand scale), and fraction of them containing a RAF (left-hand scale). The number of large chemistries decreases as the cleavage probability increases, until they are absent for $p_{cl} = 0.9$. (c) The number of RAFs present, divided between those within large chemistries and those within small chemistries. The contribution of large chemistries decreases $p_{cl}$ increases (because the probability of obtaining large chemistries decreases) while the contribution of small chemistries increases only slightly. (d) Size of RAFs relative to the chemistries they belong to: RAFs in large chemistries are as large as their host chemistries, while those in small chemistries are smaller than them

d) In both cases, there are different reactions able to lead to the same species. The "number of different catalysts" which lead to the synthesis of the same product is shown in Fig. 9 as a function of the length of the product. There are many more catalysts in the case of K-chemistry, and also the trends are very different: the ways to obtain the same product are a decreasing function of its length in the K-case, while there is a maximum value for C-chemistries.
e) As expected, in both cases the number of reactions in which a particular species appears as a product is a decreasing function of its length (Fig. 8). We can also note that, at any length, the variability in C chemistry is greater than that in K chemistry

Note also that in a C-chemistry, the number of catalyzed reactions obviously depends upon the length of the active site (as shown in Fig. 10) while in K-systems there is no active site.

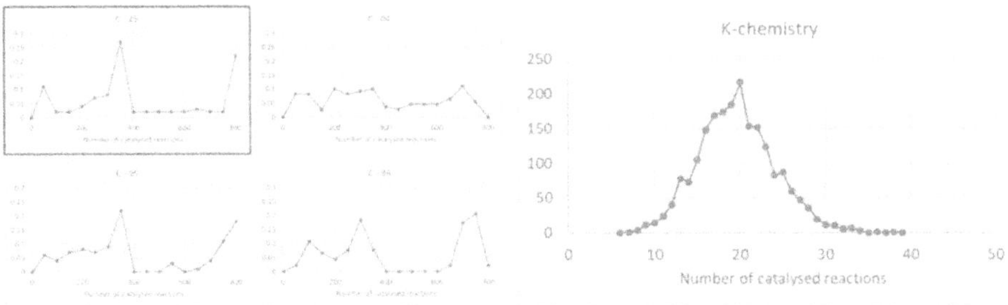


(a) (b)

**Fig. 7.** The distribution of the number of reactions catalyzed by each catalyst. (a) Four C-chemistries: the distribution is different from case to case: sometimes almost uniform, sometimes irregular with the presence of multiple peaks. The distribution of the chemistry described here is the one on the top left. (b) The distribution of the K-chemistries is always the same, and takes on a bell shape, as in the K-chemistry used here (note the different scale of the X-axis, compared to the scale of the distributions of the C-chemistries)

All the examined large C-chemistries present large RAFs: however, as the cleavage fraction increases, the probability of observing a large chemistry decreases. We repeated the same experiment with large K-chemistries, maintaining at each value of $p_{cl}$ the number of species and the average connectivity of the large C-chemistries.

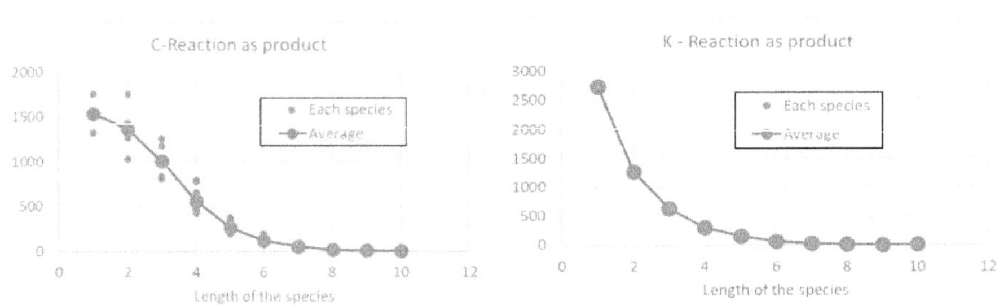


(a) (b)

**Fig. 8.** Number of reactions in which a particular species appears as a product, as a function of length. (a) C-chemistry (each species, and the average over each length). (b) K-chemistry (each species, and the average over each length). We can also note that, at least for short-length species, the variability in C-chemistry is greater than that in K-chemistry.

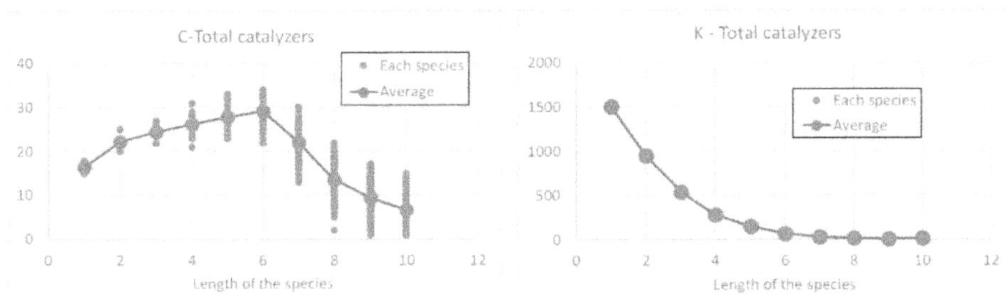


(a) (b)

**Fig. 9.** Number of catalysts contributing to the formation of a particular species, as a function of length. (a) C-chemistry (each species, and the average over each length) – it may be recalled that in this particular chemistry there are only 100 catalysts. (b) K-chemistry (each species, and the average over each length). The trends are different, as are the values. We can also note that, at any length, the variability in C chemistry is greater than that in K chemistry.

The increase in the cleavage fraction has an influence also in this context, lowering the probability that a large K-chemistry hosts a RAF (Fig. 11). A similar behavior, but in a different context, had been identified in [25].

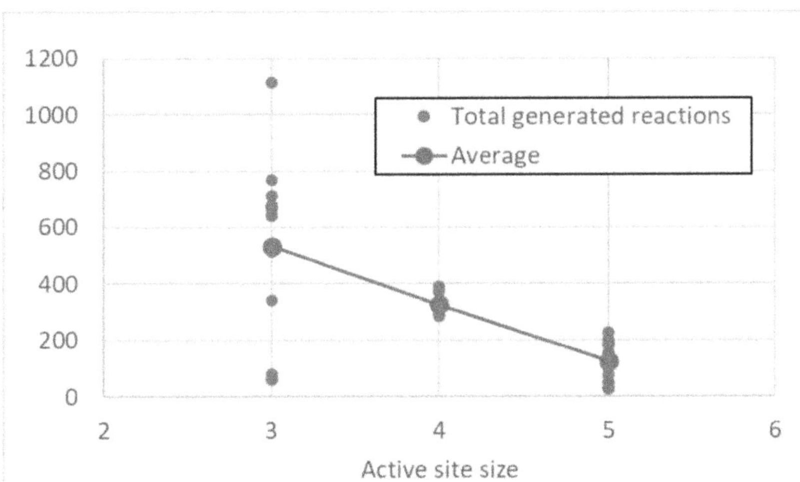


**Fig. 10.** Number of reactions generated by each catalyst in C chemistry, as the size of the active site varies: single values, and average for each length (to highlight the phenomenon, in this C-chemistry we used active site sizes in the range [3, 5]). There is no equivalent in K chemistry.

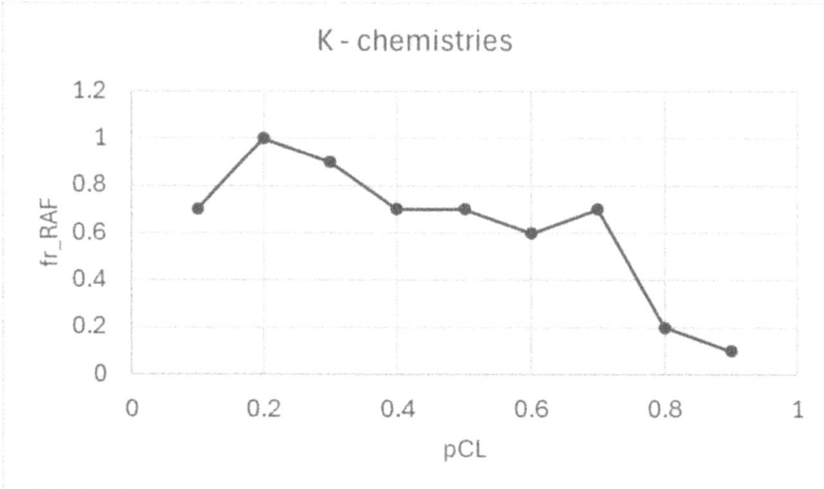


**Fig. 11.** Influence of the relative number of cleavages and condensations in K-chemistry.

## 4 RAF in Protocells

The growth of a population of protocells takes place by (i) growth of newborn protocells followed by (ii) fission of a mother cell into two daughters, further followed by their growth, etc. A sustainable growth requires that newborn protocells resemble their parents, i.e. that fission takes place exactly when the number of self-replicating molecules has doubled from the time of their birth. Which in turn requires some form of "synchronization" [11, 26] of the two processes of fission and molecular replication. In present-day cells synchronization is guaranteed by sophisticated control mechanisms whose existence cannot be hypothesized in simpler protocells, since they are the outcome of long evolutionary processes.

We have shown in previous works [11, 27–30] that such synchronization is a widespread emergent property, in particular when the kinetic equations are those of the law of mass action, and the presence of the necessary material - the "food" in the terminology of RAFs - is guaranteed from the environment.[1] However, two different cases may take place, i.e. one where there is a sustained growth of the number of protocells (which can be eventually limited by availability of resources), and another in which there is asymptotically no more duplication. While the model does not explicitly deal with cell death, this latter case may be identified with starvation.

A sustained growth of the number of protocells requires the presence of a RAF (since it allows the production of both the reactants and the necessary catalysts) in artificial chemistry, but this is not sufficient, since fission may lead to dilution and eventually to

[1] The architecture of the protocell is such that some substances can cross the membrane by passive diffusion, the flow being driven by the gradient between the internal and external concentrations. In this way, if internally the concentrations of these substances tend to differ from the external ones due to internal chemical reactions, a flow connected to the outside is activated tending to counteract this variation.

starvation. In order to study this possibility, it is necessary to adopt a truly dynamical approach to verify whether the RAFs present in the structure of the artificial chemistry ("static" RAFs) are able to allow the dynamic sustainability of the protocell.

Simulations require determining the kinetic constants of cleavage and condensation reactions, the diffusion coefficient through the membrane for substances capable of crossing it, and the kinetic constant of reactions capable of producing membrane materials.

There is a very high number of degrees of freedom, and it is not possible to derive general rules. Via simulation, some works have observed that the variation of the ratio of the typical values of the cleavage kinetic constants with respect to the typical values of the condensation kinetic constants can lead the same chemistries sometimes to synchronization and sometimes to starvation [31]. However, other works indicate that large RAFs have extreme difficulty in synchronizing, even applying different combinations of kinetic constants [11].

Given the different organization of C-chemistries (compared to that of K-chemistries), it is therefore interesting to observe what happens in the case of RAFs present within them.

In the case of large RAFs, the simulations are extremely long, so we proceeded with sampling (a small number of simulations - 10), never obtaining synchronization - the situation might therefore be similar to that existing for RAFs present in K-chemistries.

On the contrary, all the (few) RAFs present in small-sized C-chemistries (17 RAF, out of a total of 757 small chemistries) synchronize, regardless of the $p_{cl}$ value of the chemistry containing them. Their size is extremely small: there are few species (1–2 tens at most) of which a small fraction are catalysts. The core of RAFs is frequently a reciprocal catalysis situation (one species catalyzes the formation of another, and vice versa).

## 5 Conclusions

In this work we show the effects of introducing structure-dependent catalysis mechanisms into a polymer chemistry model first introduced by Kauffman in his research on the emergence of autocatalytic systems and the origin of life: in particular, we focused on situations where the action of catalysts is not very specific, a plausible scenario in the case of primitive proto-biological systems. Such introduction profoundly modifies the organization of the chemistries that can be built, and consequently the formation of autocatalytic systems within them. Compared to K-chemistries, C-chemistries have few catalysts, which catalyze a large number of reactions; the number of catalysts that lead to the formation of single chemical species of equal length can be very variable. C-chemistries tend to be small with low connectivity, or large with very high connectivity: intermediate-sized chemistries are almost entirely absent. Large-sized chemistries always have a large RAF inside them. It is possible, although not very common, for small-sized C-chemistries to have RAFs.

Despite the strong differences between C-chemistries and K-chemistries, it is interesting to note that in both models an increase in the cleavage fraction $p_{cl}$ leads to a decrease in the probability of having RAFs.

Finally, the results support the hypothesis that large RAFs have difficulty synchronizing, in both types of models. There is a probability - small but significant - that small-sized C-chemicals host RAFs, in these cases capable of synchronizing. This result deserves to be further investigated, because of its repercussions in the field of theories about the origin of life.

**Acknowledgments.** This research was partially funded by Università degli Studi di Modena e Reggio Emilia (FAR2025 project of the Department of Physics, Informatics and Mathematics).

**Disclosure of Interests.** The authors have no competing interests to declare that are relevant to the content of this article.

## References

1. Alberts, B., Johnson, A., Lewis, J., Raff, M., Roberts, K., Walter, P.: Molecular Biology of the Cell. 6th edn. Garland Science (2014)
2. Rasmussen, S., et al. (eds.): Protocells. MIT Press, Cambridge (2008)
3. Kauffman, S.A.: Autocatalytic sets of proteins. J. Theor. Biol. **119**, 1–24 (1986)
4. Bagley, R.J., Farmer, J.D., Kauffman, S.A., Packard, N.H., Perelson, A.S., Stadnyk, I.M.: Modeling adaptive biological systems. Biosystems **23**(2–3), 113–137 (1989)
5. Deamer, D.W., Fleischaker, G.R.: Origins of Life: The Central Concepts. Jones and Bartlett Publishers, Boston (1994)
6. Davies, P.: The Origin of Life. Penguin (2003)
7. Pross, A.: What is Life? How Chemistry Becomes Biology. Oxford University Press, Oxford (2012)
8. Kauffman, S.A.: The Origins of Order: Self-Organization and Selection in Evolution. Oxford University Press (1993)
9. Kauffman, S.A.: At Home in the Universe: The Search for the Laws of Self-Organization and Complexity. Oxford University Press (1995)
10. Serra, R., Carletti, T., Poli, I.: Synchronization phenomena in surface-reaction models of protocells. Artif. Life **13**, 123–138 (2007)
11. Serra, R., Villani, M.: Modelling Protocells. The Emergent Synchronization of Reproduction and Molecular Replication. Springer, Dordrecht (2017)
12. Bagley, R.J., Farmer, J.D.: Spontaneous emergence of a metabolism. In: Langton, C., Taylor, C., Farmer, J.D., Rasmussen, S. (eds.) Artificial Life II. Santa Fe Institute Studies in the Sciences of Complexity, vol. X, pp. 93–140. Addison-Wesley, Redwood City (1991)
13. Filisetti, A., et al.: A stochastic model of autocatalytic reaction networks. Theory Biosci. **131**(2), 85–93 (2011)
14. Hordijk, W., Steel, M.: Detecting autocatalytic, self-sustaining sets in chemical reaction systems. J. Theor. Biol. **227**, 451–461 (2004)
15. Hordijk, W., Hein, J., Steel, M.: Autocatalytic sets and the origin of life. Entropy **12**, 1733–1742 (2010)
16. Hordijk, W., Wills, P.R., Steel, M.: Autocatalytic sets and biological specificity. Bull. Math. Biol. **76**, 201–224 (2014)
17. Shapiro, R.: Origins: A Skeptic's Guide to the Creation of Life on Earth. Summit Books (1986)
18. Darwin, C.: Letter to Joseph Dalton Hooker, 1 February 1871. The Correspondence of Charles Darwin, vol. 19, p. 215. Cambridge University Press (1871)
19. Oparin, A.I.: The Origin of Life on Earth, 3rd edn. Oliver and Boyd, Edinburgh (1957)

20. Haldane, J.B.S.: Origin of Life. The Rationalist Annual, pp. 3–10 (1929)
21. Könnyű, B., Czárán, T.: The evolution of enzyme specificity in the metabolic replicator model of prebiotic evolution. PLoS ONE **6**(6), e20931 (2011)
22. de Graaf, R., De Decker, Y., Sojo, V., Hudson, R.: Quantifying catalysis at the origin of life chem. Eur. J. **29**, e202301447 (2023)
23. Wolfenden, R., Snider, M.J.: The depth of chemical time and the power of enzymes as catalysts. Acc. Chem. Res. **34**(12), 938–945 (2001)
24. Berg, J.M., Tymoczko, J.L., Stryer, L.: Biochemistry, 8th edn. W.H. Freeman (2015)
25. Füchslin, R.M., Filisetti, A., Serra, R., Villani, M., Delucrezia, D., Poli, I.: Dynamical stability of autocatalytic sets. In: Fellermann, H., et al. (eds.) Artificial Life XII, Proceedings of the Twelfth International Conference on the Synthesis and Simulation of Living Systems. MIT Press, Cambridge (2010)
26. Serra, R., Villani, M.: Sustainable growth and synchronization in protocell models. Life **9**, 68 (2019)
27. Carletti, T., Serra, R., Poli, I., Villani, M., Filisetti, A.: Sufficient conditions for emergent synchronization in protocell models. J. Theor. Biol. **254**, 741–751 (2008)
28. Filisetti, A., et al.: A stochastic model of the emergence of autocatalytic cycles. J. Syst. Chem. **2**(2) (2011). https://doi.org/10.1186/1759-2208-2-2
29. Villani, M., Filisetti, A., Graudenzi, A., Damiani, C., Carletti, T., Serra, R.: Growth and division in a dynamic protocell model. Life **4**, 837–864 (2014)
30. Villani, M., Alboresi, E., Serra, R.: Models of protocells undergoing asymmetrical division. Entropy **26**, 281 (2024)
31. Villani, M., Serra, R.: Protocells either synchronize or starve. Entropy **27**, 154 (2025)

# Thermodynamics of Biological Switches

Roger D. Jones[1,2(✉)], Achille Giacometti[1], and Alan M. Jones[2,3]

[1] European Centre for Living Technology, Department of Molecular Sciences and Nanosystems, Ca' Foscari, University of Venice, Venice, Italy
RogerDJonesPhD@gmail.com

[2] Department of Biology, University of North Carolina at Chapel Hill, Chapel Hill, USA

[3] Department of Pharmacology, University of North Carolina at Chapel Hill, Chapel Hill, USA

**Abstract.** We derive a formulation of the First Law of nonequilibrium thermodynamics for biological information-processing systems by partitioning entropy in the Second Law into microscopic and mesoscopic components and by assuming that natural selection promotes optimal information processing and transmission. The resulting framework demonstrates how mesoscopic information-based subsystems can attain non-equilibrium steady states (NESS) sustained by external energy and entropy fluxes, such as those generated by ATP/ADP imbalances *in vivo*. Moreover, mesoscopic systems may reach NESS before microscopic subsystems, leading to ordered structures in entropy flow analogous to eddies in a moving stream.

**Keywords:** biological switches · nonequilibrium thermodynamics · Second Law of Thermodynamics · mutual information · entropy · free energy

## 1 Introduction

The universe is subject to the Second Law of Thermodynamics and is irreversibly moving toward heat death, a state of maximal disorder in which information and structure vanish [15]. However, within this trajectory toward uniformity, life has emerged as a striking counterexample: an organized, information-processing phenomenon inseparably embedded in a turbulent physical environment. To persist, living systems must continually adapt, and adaptation depends on the acquisition, transmission, and transformation of information. In this sense, biology can be understood as a form of computation.

At the molecular scale, this computation is carried out by regulatory processes that function analogously to switches [6,8,9]. These mechanisms, fueled by external energy sources such as sunlight or nutrient-derived metabolites, govern cellular behavior through precise biochemical reactions. The phosphorylation/dephosphorylation cycle (PdPC) (see Fig. 1) exemplifies this mechanism


M. Bianchini et al. (Eds.): WIVACE 2025, CCIS 3001, pp. 106–114, 2027.
https://doi.org/10.1007/978-3-032-33185-4_8

[14]. These reversible reactions, typically occurring in amino acids within proteins, act as decision points that direct cellular outcomes under a wide range of conditions. Other classes of molecular switches are widespread, yet share the defining characteristics of this cycle [14]. Through their collective action, molecular switches dynamically regulate physiological functions, including immune responses, behavior, and metabolism. Life's adaptability thus emerges from the orchestrated interplay of countless such reactions, which continuously compute and recalibrate biological information.

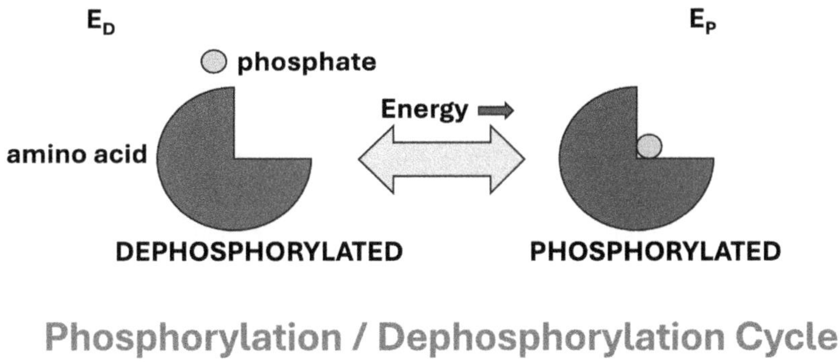


**Fig. 1.** An example of a simple chemical reaction commonly used as a biological switch. Typically, energy input is required to phosphorylate an amino acid. ATP imbalance drives the energy $E_D$ of the dephosphorylated state to the higher energy $E_P$ of the phosphorylated state. The switch lives in a protein matrix that, through conformational changes, alter the energy relations between phosphorylated and dephosphorylated states. Whether a site is phosphorylated or not can determine intracellular response to the switch state.

From a broader perspective, life represents a process by which order is carved from thermodynamic chaos. By coarse-graining innumerable microscopic states into stable functional patterns shaped by natural selection, biological systems maintain complexity despite entropic pressures. Sustained by external energy fluxes within an otherwise equilibrating universe, these systems manipulate information to create mesostates (nanometer length scale) that respond flexibly to environmental stimuli. In this way, life persists as a transient but resilient island of order, demonstrating how matter and energy can generate computational architectures capable of resisting the universal drift toward randomness.

Simple molecular switches, such as those illustrated in Fig. 1, lend themselves to analysis using statistical-physics methods similar to those described in the monograph by Seifert [16]. Here, we apply statistical physics to a biological information-processing system immersed in a thermal bath and coupled to a protein matrix capable of modulating the switch's energy levels. By incorporating Shannon's concept of mutual information [17] into the Second Law of Thermodynamics, we show that such systems can maintain functionality over extended timescales and operate at an extremum of information processing. This perspective further suggests a generalization of the First Law of Thermodynamics in which information itself contributes to the generation of heat.

This study complements a parallel investigation [7] that derives the statistical ensemble for biological switches powered by dedicated energy sources. That framework, termed the Biological Ensemble, generalizes the Gibbs ensemble to describe open biological systems and shows that such switches can occupy quasistable states corresponding to on, off, and inactive configurations.

## 2 The Model Switch

### 2.1 Microstates and Mesostates

The model switch considered here, exemplified by the phosphorylation/dephosphorylation cycle (PdPC) shown in Fig. 1, consists of atoms and molecules whose precise configuration at any instant defines a microstate $x \in X$, where $X$ is the set of all possible microstates. The task of statistical mechanics is to determine the probability of a given microstate $x$ based on mesoscopic knowledge of the system and externally imposed constraints. In equilibrium statistical mechanics, this is accomplished by the Canonical Ensemble [15], from which the laws of equilibrium thermodynamics follow. In that case, the average energy of the microstates defines the temperature as a fundamental mesoscopic variable.

From a functional perspective, the switch is described by a mesostate $y$, where $y = (1, 0)$ denotes the dephosphorylated form and $y = (0, 1)$ denotes the phosphorylated form. Each mesostate comprises many underlying microstates; for example, numerous microscopic configurations are consistent with phosphorylation. Thus, the space of microstates can be partitioned according to their associated mesostates, and each mesostate $y$ is assigned a probability $\Pr(y)$. A rigorous treatment of these ideas is given in [11] and [12].

The switch is embedded in a thermal bath of temperature $T$ and within a protein matrix whose configurations, determined by ligand binding, modulate the phosphorylation state. Both the bath and the protein matrix can themselves be described in terms of micro- and mesostates. Together, these components make up a mesoscopic information system that must be stable, programmable, and capable of transmitting information to other systems that generate biological outcomes. The heat bath mediates thermal exchange with this system.

Switches dissipate heat when they cycle through on and off states [10], and functional information processing requires the input of external energy. The fact that information generates heat upon its erasure is known as Landauer's Principle [2].

For the PdPC, this energy is supplied by a nonequilibrium chemical flux $J$ arising from imbalances in ATP and ADP concentrations, though other nucleotides may play similar roles in different systems. Unlike equilibrium systems, biological information-processing systems are open and constrained by such fluxes, rendering the Canonical Ensemble of equilibrium thermodynamics inapplicable. Instead, the recently proposed Biological Ensemble [7] generalizes the Canonical Ensemble to these open systems, yielding thermodynamic consequences distinct from those of equilibrium ensembles, including explicit treatment of heat dissipation from information processing [10].

### 2.2 Natural Selection and Information

We postulate that natural selection favors genetic architectures that enhance the efficiency of information processing and transformation. This efficiency improves the ability of organisms to adapt to fluctuating environments, thereby prolonging survival and increasing the persistence of genetic material across generations. For single switch, this postulate leads to the observation that the switch can exist in up to three quasistable states for any specific values of the phosphate bonding energy [7]. This behavior has been observed in impedance measurements of photo-active ligands interacting with transmembrane receptors [19].

We adopt Shannon's mutual information [17] as a natural measure of biological information, since it follows from five fundamental axioms: non-negativity, symmetry, chain rule, data processing inequality, and additivity [3]. Relaxing one or more of these axioms produces alternative measures, but Shannon's formulation remains the most widely applicable. For a mesoscopic system that transmits information from one storage state $y \in Y$ to another $y' \in Y$, the transferred information $I_M$ is given by

$$I_M = \sum_{y \in Y} \sum_{y' \in Y} \Pr(y, y') \ln \left( \frac{\Pr(y, y')}{\Pr(y) \Pr(y')} \right) \tag{1}$$

where $\Pr(y)$ and $\Pr(y')$ denote the probabilities of the respective mesostates and $\Pr(y, y')$ the joint probability distribution. For a switch such as the PdPC (Fig. 1), the set of mesostates can be written as

$$Y = (d, p), \quad p \in (0, 1), \; d \in (0, 1) \tag{2}$$

where $d$ represents the dephosphorylated state and $p$ the phosphorylated state.

If natural selection acts to maximize information processing efficiency, the principle can be formalized as

$$dI_M \geq 0, \tag{3}$$

implying that evolutionary dynamics preserve or increase the mutual information transmitted by biological switches.

### 2.3 Second Law of Thermodynamics

The path toward thermodynamics of biological systems proceeds through the Second Law of Thermodynamics, which holds in both equilibrium and nonequilibrium regimes. Adopting Jaynes' informational formulation of entropy in the Second Law [5], we derive a corresponding First Law that describes the flow of energy among information, work, and the thermal bath.

The Gibbs-Shannon entropy $S$ of a classical physical system composed of a set $X$ of microstates $x \in X$ is (see [5])

$$S = -\sum_{x \in X} \Pr(x) \ln \Pr(x) \tag{4}$$

where $\Pr(x)$ is the probability that microstate $x$ occurs.

The Second Law of Thermodynamics can be stated as

$$dS \geq 0 \tag{5}$$

The fact that information processing generates heat implies that when the information of a mesosystem increases, the entropy of the microstates increases.

The entropy can then be restated as

$$S = S_m - I_M \tag{6}$$

where $S_m$ is the entropy of microscopic states reduced by the degrees of freedom lost to the constraints on the information $I_M$.

$$S_m = \sum_{x \in X} \sum_{y \in Y} \sum_{y' \in Y} \Pr(x, y, y') \ \ln \left( \frac{\Pr(y, y')}{\Pr(x) \Pr(y) \Pr(y')} \right) \tag{7}$$

The minus sign in Eq. 6 is chosen in order to be consistent with Landauer's Principle. Incorporation of finite information $I_M$ in Eq. 7 into microscopic entropy $S_m$ increases the entropy in microscopic states. This is a consequence of the non-negativity of information flow $I_M$.

The Second Law becomes [13,16]

$$dS = dS_m - dI_M \geq 0 \tag{8}$$

and

$$dS_m \geq dI_M \tag{9}$$

which states that the heating of the microstates is faster or as fast as information is generated. Equation 7 defines the entropy for the microstates in the presence of controlled mesoscopic information $I_M$.

### 2.4 Creation and Maintenance of Information

Mesoscopic information $I_M$ is generated and maintained by externally driven chemical flux $J$ [18]. This flux can be expressed in terms of probability flow as

$$J = \Pr(y|y') \Pr(y') = \Pr(y'|y) \Pr(y) = \Pr(y, y') \tag{10}$$

where $\Pr(y|y')$ and $\Pr(y'|y)$ are the transition probabilities between the mesostates $y$ and $y'$. In steady state, the forward flux from $y'$ to $y$ is equal to the reverse flux from $y$ to $y'$. For $y \neq y'$, Eq. 10 is Bayes theorem with $J = \Pr(y, y')$

[4]. This implies that the information can be completely specified with knowledge of $J$, $\Pr(y)$, and $\Pr(y')$. We can think of $J$ as an externally determined control parameter and take this quantity as given by external conditions.

It is important to note that Eq. 10 does not imply microscopic reversibility of the mesosystem. The system need not follow the same trajectory in the transition from $y$ to $y'$ as in the reverse transition from $y'$ to $y$. The forward and reverse pathways may differ, forming a closed reaction cycle. This distinction is essential for biological switches such as PdPCs, in which the forward and reverse reactions are catalyzed by different enzymes, typically a kinase and a phosphatase, respectively. Such catalytic asymmetry enables enzymatic control over the state of the switch [7].

Since $I_M$ can be expressed in terms of ATP-driven flux $J$, it can be maximized by varying probabilities $\Pr(y)$ and $\Pr(y')$ while keeping $J$ constant. This optimization yields quasistable states that correspond to the possible configurations of the switch [7]. These quasistable states represent local maxima of information subject to the nonequilibrium steady-state (NESS) condition

$$dI_M = 0. \tag{11}$$

It follows from Eq. 8 that all entropy production occurs at the microscopic level,

$$dS = dS_m \geq 0, \tag{12}$$

while the mesoscopic information states remain in NESS.

## 3 First Law of Thermodynamics

In the absence of work exchanged with the external environment, the First Law of Thermodynamics can be written using the Clausius relation between energy and entropy as [15]

$$dU = T\,dS - p\,dV + dW \tag{13}$$

where $U$ is the total energy of the system, $T$ is the constant temperature of the microstates, $p$ is the pressure, $V$ is the volume, and $dW$ is the work carried out on the system other than $p\,dV$ work.

Defining the Biological Free Energy $B$ as

$$B = U + pV - TS_m = H - TS_m \tag{14}$$

where $H$ is the enthalpy, we obtain

$$dB = V\,dp \;-\; S_m\,dT \;-\; T\,dI_M \;+\; dW \tag{15}$$

which, for constant temperature and pressure, becomes

$$dB = -T\,dI_M + dW \tag{16}$$

which shows that maximization of mesoscopic information $I_M$ in the absence of external work on the mesosystem corresponds to minimization of Biological

Free Energy $B$. This condition is achieved when the mesoscopic system reaches a NESS ($dI_M = 0,\ dW = 0$), which may occur well before the total entropy is maximized. The timescales of entropy maximization in the universe are much longer than the timescales of natural selection.

The molecular switches reside within proteins capable of performing mechanical work $dW$ on the switches through conformational changes in their tertiary structure. These conformational transitions are typically triggered by the binding and dissociation of ligands, coupling the chemical free energy to information processing within the protein. Consequently, the protein conformation determines or at least modulates the information content of the switch. The formation of a switch is influenced by the work generated as the protein matrix changes conformation, while the erasure of information is accompanied by the dissipation of heat into the surrounding microstates, consistent with Landauer's Principle [2]. Thus, the creation and maintenance of mesoscopic information represent a dynamic partnership between the external flux $J$, which drives energy through the mesosystem, and the mechanical work performed by the protein matrix on the information-bearing components. This framework unites thermodynamic, informational, and biological perspectives on cellular regulation.

Together, Eqs. 7, 14, and 16 generalize the Gibbs free energy from equilibrium systems to biological switches operating in NESS.

## 4 Discussion

In biological information-processing systems, energy and entropy can be partitioned into microscopic and mesoscopic components. Introducing the concept of Biological Free Energy demonstrates that its minimization corresponds to the maximization of mesoscopic information. This condition defines a nonequilibrium steady state (NESS), which can be established well before total entropy is maximized. In this way, the framework generalizes the Gibbs free energy—traditionally restricted to equilibrium—to describe the dynamics of biological switches operating far from equilibrium.

Building on the separation of entropy in the Second Law, and assuming that natural selection favors optimal information processing and transmission, we derived expressions for the First Law of nonequilibrium thermodynamics in biological information systems (Eqs. 14, 16, and 7). The analysis assumes thermal equilibrium at the level of microstates, while mesoscopic states remain out of chemical equilibrium.

The mesoscopic information-processing subsystem reaches a NESS driven by external energy fluxes, such as ATP/ADP cycling. This steady state can emerge prior to microscopic equilibrium, generating structured entropy flows analogous to eddies in a moving fluid. On evolutionary timescales, life itself can be viewed as a persistent nonequilibrium phenomenon, sustained far from thermodynamic equilibrium long before universal entropy reaches its maximum.

It is important to emphasize that the present treatment does not require mesoscale information systems to be microscopically reversible, nor are the mesostates of biological switches assumed to be in or near thermal equilibrium, although the surrounding heat bath is considered equilibrated. The only requirement is that the switch operates in an NESS, where the information is located at an extremum.

The absence of microscopic reversibility implies that phosphorylation and dephosphorylation proceed along distinct reaction pathways, a necessity *in vivo*, where kinases catalyze phosphorylation and phosphatases drive dephosphorylation. Consequently, these reactions are thermodynamically coupled yet mechanistically distinct. The informational mesosystem thus functions far from thermochemical equilibrium. The probability distributions of switch states differ qualitatively from those of equilibrium systems; for a given phosphate-bonding energy, the system may occupy up to three quasistable states, in contrast to the single state permitted under equilibrium conditions [7].

The model of a biological switch regulated by a measurable chemical flux $J$ can be generalized to systems in which both total energy input and heat output are characterized by $Q = Jq$, where $Q$ represents the total rate of energy input from ATP hydrolysis and $q$ denotes the heat dissipated per switching cycle between on and off states. In networks of multiple switches, variations in heat dissipation among individual elements may determine the sequence of activation. Such thermodynamic coupling likely plays a critical role in switch regulation and will be explored further through combined theoretical and experimental studies.

Historically, stochastic thermodynamics [13,16] has addressed the resolution of Loschmidt's paradox, which questions the compatibility of microscopic reversibility with the Second Law of Thermodynamics [1]. The time dependence of transient relaxation on a steady state, typically assumed to be equilibrium, has been central to this effort, leading to fluctuation theorems that describe the mesoscopic nature of these transients [20].

In contrast, the present treatment does not restrict the steady state to equilibrium; instead, it explicitly requires a NESS. Furthermore, we do not address the transient approach to NESS. The relevant notion of reversibility here is not microscopic, but logical reversibility [2]. If no work is performed on the mesosystem ($dW = 0$) by the surrounding protein matrix, then NESS requires that no change occurs in the information content of the mesosystem ($dI_M = 0$). This condition implies that the mesosystem is logically reversible, though not necessarily microscopically reversible.

The PdPC model developed here can easily be generalized to a wide range of biological switches, including networks of multiple interacting switches and information systems operating under diverse constraints. The entropy change of the microscopic states may differ across mesoscopic configurations, reflecting the dependence of molecular dynamics on higher-level organization. Transitions between on and off states can occur through conformational changes in the surrounding protein matrix, which functions as an external constraint shaping the accessible mesostates.

**Acknowledgements.** Research by AG is supported by MIUR PRIN-COFIN2022 grant 2022JWAF7Y.

**Disclosure of Interests.** The authors have no competing interests to declare that are relevant to the content of this article.

## References


1. Binder, P.: The reversibility paradox: role of the velocity reversal step. Int. J. Theor. Phys. **62**(9), 200 (2023)
2. Chattopadhyay, P., Misra, A., Pandit, T., Paul, G.: Landauer principle and thermodynamics of computation. Rep. Prog. Phys. (2025)
3. Cover, T.M.: Elements of Information Theory. Wiley (1999)
4. Grover, J.: A literature review of bayes' theorem and Bayesian belief networks (BBN). In: Strategic Economic Decision-Making: Using Bayesian Belief Networks to Solve Complex Problems, pp. 11–27 (2012)
5. Jaynes, E.T.: Information theory and statistical mechanics. Phys. Rev. **106**(4), 620 (1957)
6. Jones, R.D.: Information transmission in g protein-coupled receptors. Int. J. Mol. Sci. **25**(3), 1621 (2024)
7. Jones, R.D., Giacometti, A., Jones, A.M.: Information transmission and processing in g-protein-coupled-receptor complexes. arXiv preprint arXiv:2508.11039 (2025)
8. Jones, R.D., Jones, A.M.: Model of ligand-triggered information transmission in g-protein coupled receptor complexes. Front. Endocrinol. **14**, 879 (2023)
9. Jones, R.D., Jones, A.M.: A proposed mechanism for in vivo programming transmembrane receptors. In: Italian Workshop on Artificial Life and Evolutionary Computation, pp. 123–137. Springer (2023)
10. Landauer, R.: The physical nature of information. Phys. Lett. A **217**(4–5), 188–193 (1996)
11. Marletto, C.: Constructor theory of probability. Proc. Roy. Soc. A Math. Phys. Eng. Sci. **472**(2192), 20150883 (2016)
12. Marletto, C.: Constructor theory of thermodynamics. arXiv preprint arXiv:1608.02625 (2016)
13. Parrondo, J.M., Horowitz, J.M., Sagawa, T.: Thermodynamics of information. Nat. Phys. **11**(2), 131–139 (2015)
14. Qian, H.: Phosphorylation energy hypothesis: open chemical systems and their biological functions. Annu. Rev. Phys. Chem. **58**, 113–142 (2007)
15. Reif, F.: Fundamentals of Statistical and Thermal Physics. Waveland Press (2009)
16. Seifert, U.: Stochastic thermodynamics: principles and perspectives. Eur. Phys. J. B **64**(3), 423–431 (2008)
17. Shannon, C., Weaver, W.: The mathematical theory of communication, (first published in 1949). Urbana University of Illinois Press (1963)
18. Snoke, D.W.: The crucial role of thermodynamic gates in living systems. BioCosmos **4**(1), 20240004 (2024)
19. Wirth, U., et al.: Monitoring the reversibility of GPCR signaling by combining photochromic ligands with label-free impedance analysis. Angew. Chem. **135**(21), e202215547 (2023)
20. Yang, H., Ge, H.: Fluctuation theorems and thermodynamic inequalities for nonequilibrium processes stopped at stochastic times. Phys. Rev. E **108**(5), L052104 (2023)

# Complex System Exploration with Interactive Human Guidance

Bastien Morel[1(✉)], Clément Moulin-Frier[2], and Pascal Barla[1]

[1] Univ. Bordeaux, CNRS, Bordeaux INP, Inria, LaBRI, UMR 5800, 33400 Talence, France
{bastien.morel,pascal.barla}@inria.fr
[2] Flowers AI & CogSci Lab, Inria, Talence, France
clement.moulin-Frier@inria.fr

**Abstract.** The diversity of patterns that emerge from complex systems motivates their use for scientific or artistic purposes. When exploring these systems, the challenges faced are the size of the parameter space and the strongly non-linear mapping between parameters and emerging patterns. In addition, artists and scientists who explore complex systems do so with an expectation of particular patterns. Taking these expectations into account adds a new set of challenges, which the exploration process must address. We provide design choices and their implementation to address these challenges; enabling the maximization of the diversity of patterns discovered in the user's region of interest – which we call the constrained diversity – in a sample-efficient manner. The region of interest is expressed in the form of explicit constraints. These constraints are formulated by the user in a system-agnostic way, and their addition enables interactive system exploration leading to constrained diversity, while maintaining global diversity.

**Keywords:** Complex systems · automated discovery · user interaction

## 1 Introduction

Complex systems are prone to the self-organization of spatio-temporal patterns from the local interaction of their constituents. In nature, many patterns result from the dynamics of such complex systems [1], which has motivated the artificial life community to rely on computer simulations in order to understand and explore such phenomena [4], as models of natural phenomena, or for generative arts. Taking advantage of complex system, requires to efficiently explore the space of patterns they can generate. Exploring an artificial complex system involves sampling the parameter space that defines it, with the aim of finding a specific pattern or discovering a set of diverse new patterns. Searching for patterns produced by a complex system requires dealing with excessively large, potentially open-ended search spaces [5], combined with a strongly non-linear mapping between system's parameters and system's behaviors. Random sampling in parameter space is inefficient compared to machine learning methods


M. Bianchini et al. (Eds.): WIVACE 2025, CCIS 3001, pp. 115–128, 2027.
https://doi.org/10.1007/978-3-032-33185-4_9

based on optimization or diversity. In particular, intrinsically motivated goal exploration process (IMGEP) [22] tends to maximize the diversity of discovered patterns in behavior space, i.e., a representation space characterizing the set of possible patterns in a given system (more details are given in Sect. 4).

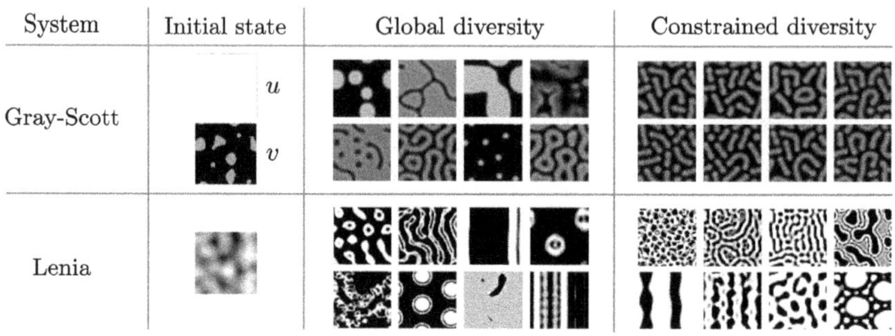


**Fig. 1. Our method explores the parameter space of complex systems to discover patterns that belong to the user's region of interest.** Compared to global diversity search, our constrained diversity search approach allows the exploration process to be guided toward user-specific patterns. In this example, the user expects to obtain images with a dark area proportion belonging to $[0.3, 0.4]$.

In addition to a large/open-ended search space, exploration can take place with particular expectations in the user's mind, and feedback is needed to provide them. Adding a human to the exploration loop forces the exploration process to take user constraints into account. The goal is then to achieve what we call a *constrained diversity* of patterns (see Fig. 1). Indeed, of all the patterns that can be generated by a complex system, the user may only be interested in a subset, which we term a Region Of Interest (ROI). But as the system is explored, this ROI may vary, as the user takes advantage of the serendipity offered by exploration. In addition, not all points in the ROI may be accessible to the complex system.

In this work, our goal is to put the human right into the loop of complex system exploration. An overview of our approach is given in Fig. 2. In particular, we seek *interactive* feedback, meaning the exploration loop should be fast enough for the user to adapt their ROI at each iteration. In addition, we want our method to be readily usable without any prior knowledge on the system behavior. We expose design choices that allow for such an interactive exploration process of complex systems, taking into account potentially varying ROI. We implement those design choices through a modified IMGEP algorithm, and we show the ability of our implementation to maximize constrained diversity through quantitative and qualitative results.

Lastly, we discuss merits and flaws of those design choices and we identify avenues of future work.

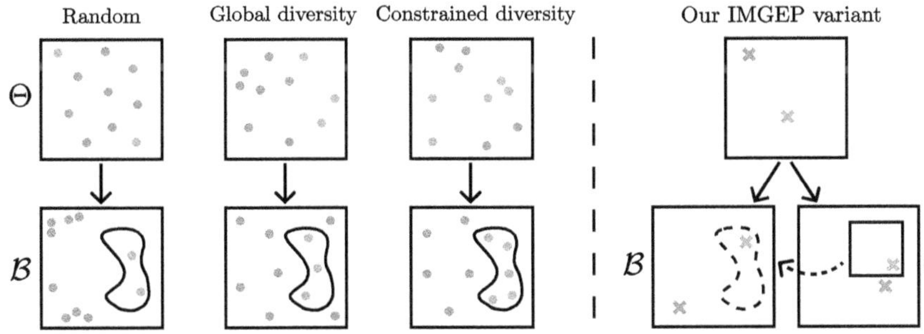


**Fig. 2. Overview of our approach.** Our goal is to define a procedure for exploring the parameter space $\Theta$ of a complex system, taking into account the target user's ROI (closed region) in a behavior space $\mathcal{B}$. A random sampling in $\Theta$ leads to a poor diversity in $\mathcal{B}$. Diversity-based methods usually achieve a high global diversity. However, most of the samples end up outside the ROI. Our goal is to find a sampling method that maximizes constrained diversity (i.e., resulting in more samples inside the user's ROI and whose diversity is maximized). To this end, we introduce an IMGEP variant that relies on an additional constraint space $\mathcal{C}$ in which the ROI is more easily expressible than in $\mathcal{B}$.

## 2 Related Works

Optimization-based methods, such as evolutionary algorithms [16] or gradient descent [17], can be used to find parameters that lead to a pattern with *a priori* optimal characteristics. What characteristics are considered as optimal has to be defined by the user in the form of a scalar objective function of the parameter space. Optimization-based methods are subject to standard issues in optimization, in particular deception in the presence of strong local optima.

Diversity-based methods aim to find sets of parameters that lead to a diversity of patterns, in a so-called behavior space. Intrinsically motivated goal exploration process (IMGEP) [7,22] belongs to this class. However, defining a behavior space a priori can potentially limit the global diversity of the discovered patterns. To circumvent this problem Etcheverry et al. [6] introduce the use of a hierarchy of behavioral spaces, a hierarchy built using a *variational autoencoder* (VAE) [13], which implies a high calculation cost.

In order to benefit from the advantages of diversity-based and optimization-base methods, hybrid methods have focused on finding parameters that lead to a diversity of patterns with a priori optimal characteristics. MAP-Elites [18] is a method that tends to maintaining a population (set of parameters) with the highest fitness scores. The behavioral space is divided into sub-sections, and the algorithm seeks to find for each sub-section the parameters with the highest fitness scores. Like IMGEP, the result of MAP-Elite depends on the a priori formulation of the behavior space. AURORA [9] relaxes this constraint and allows the use of a behavior space that is not hand-coded. IMGEP instances can also be defined to benefit from gradient descent. Hamon et al. [10] take advantages of curriculum learning capabilities and enable the discovery of increasingly complex behaviors through the use of an optimization process inside the parameter sampling strategy.

In this work, we aim to place the user in the exploration loop as a guide, without any a priori knowledge of the system or the mapping of parameters to behavior. IMGEP-HOLMES [6] enables such user guidance, by considering a hierarchy of behavior spaces and allowing the user to regularly rate each of them, thereby biasing future exploration with user feedback. However, the implicit definition of the ROI (through user rating) and the computational cost of updating the hierarchy implies a tradeoff between sample efficiency and ROI exploration speed. OMNI [26] enables open-ended learning in a vast research space, with the "interesting" tasks to be learned determined by a foundation model. The ROI is therefore modeled by a foundation model, but is difficult to customize and is built from existing data. ASAL [14] also proposes the use of foundation models for automated discovery, notably to maximize global diversity (illuminate the behavior space) or find parameters corresponding to a target (image or prompt). The core of these methods relies on having a human agent give instructions on the direction the discovery should take. However, these methods are positioned in a non-interactive context, not allowing rapid redirection of the search when the user changes his ROI.

## 3 Design Choices

In this section, we give the desiderata induced by the interactive addition of the human in the loop, followed by design choices made to satisfy these desiderata.

### 3.1 Desiderata

**D1 Constrained diversity** Matching the user's ROI to the system's behavior space is not always feasible. Hence, to ensure that the ROI has been effectively explored thoroughly, it is necessary that the diversity found within the ROI at the end of the exploration be maximal; we call this diversity, the constrained diversity.

**D2 System-agnostic** We place exploration in a context in which the user does not necessarily have any prior knowledge of the parameter-behavior mapping of the system being explored. The exploration process must therefore be agnostic to the complex system being explored, and must not be based on data from the system prior to exploration, aside from the type of generated pattern.

**D3 User feedback** As the user can change his mind and adjust his ROI during exploration, at any point during the exploration, the user must be able to steer the exploration in a direction that suits him best.

**D4 Interactive exploration** The fact that the user can give feedback to adapt the algorithm's behavior implies a fast response time for the algorithm, so that the user can determine whether his guidance is relevant or not. This means that the exploration loop must be executed in interactive time.

### 3.2 Choices

**C1 Explicit constraints** We let users provide feedback on exploration in the form of explicit constraints on pattern properties (D1). Sampled patterns are thus labelled as inliers if they lie within the ROI, and outliers otherwise, independently of the choice of system (D2). Users may also change their mind through modifications of the ROI during exploration (D3).

**C2 Maintaining global diversity** We propose to not only maximize constrained diversity, but also global diversity, with two objectives in mind. First, it should give the user a better overview of the patterns attainable by the system, as this is not known in advance (D2). Second, having a diverse set of samples (inliers or outliers) may also turn out to be useful whenever the user changes their ROI during the exploration process (D4).

**C3 Non-specific behavior space** We construct the behavior space from a set of non-system-dependent features, which has two benefits : it does not require a priori knowledge on the system (D2); neither does it require any costly a posteriori analysis of system outputs, such as VAE training (D4). This often results in a high-dimensional behavior space, which requires special care.

## 4 Background

### 4.1 Complex System

Formally, in this paper, a system is a tuple $\mathcal{S} = (\mathcal{A}, U, \Theta)$, where $A$ is the state space, $U$ the update rule and $\Theta$ the parameter space. Following Ladyman et al. [15], a *complex* system is a system composed of many entities that interact with one another.

We consider $n_s$ entities each having a set of states of dimension $d_s$. The $n_s \times d_s$ dimensional space $\mathcal{A}$ corresponds to the state space of the system. At each time step, the state of the system is updated following the update rule $U : \mathcal{A} \times \Theta \to \mathcal{A}$. The update rule is parameterised by the current state of the system $A_t$ and a parameters $\theta \in \Theta$.

A rollout $R_{0\to T}$ of a system $\mathcal{S}$, with the initial state $A_0 \in \mathcal{A}$ and parameterised by $\theta \in \Theta$, is the sequence of states $R_{0\to T}(\mathcal{S}, A_0, \theta) = A_0 \to A_1 \to ... \to A_T$ with each obtained by the iterative application of the update rule $A_t = U(A_{t-1}, \theta)$.

An observation of a rollout is obtained by applying a function $O : \mathcal{A}^T \to \mathcal{O}$, where $\mathcal{O}$ is the observation space. In the general case, observation is defined as the sequence of states obtained by the rollout (i.e., $\mathcal{O} := \mathcal{A}^T$). In our work, we use the word *observation* and *result* interchangeably, since observation is what is shown to the user as the result of a rollout.

### 4.2 Intrinsically Motivated Goal Exploration Process (IMGEP)

In this article, the initial state is not varying from one rollout to another of the same system and the result proposed to the user corresponds only to the last state of the rollout, therefore for sake of clarity $R_{0\to T}(\mathcal{S}, A_0, \theta) := R_{0\to T}(\mathcal{S}, \theta)$ and $O(R_{0\to T}(\mathcal{S}, \theta)) := A_T$.

We implement our design choices through an Intrinsically Motivated Goal Exploration Process (IMGEP). Originally developed to model curiosity-driven learning in robotics [7], IMGEP has been transposed to complex system's diversity search [5].

In this formulation, the aim of an IMGEP is to maximize diversity found in the behavior space $\mathcal{B}$ (a dimensionaly-reduced space that summarizes observations and emphasizes diversity), in a sample-efficient manner. To do this, an IMGEP carries out a sequence of experiments, each corresponding to a rollout of the system $R_{0\to T}(\mathcal{S}, \theta_i)$, with $i$ the $i$-th experiment. The parameter $\theta_i$ of all experiments, their corresponding results $o_i$ and behavior $b_i$ are stored in a history $\mathcal{H} = \{(\theta_i, o_i, b_i), \forall i \in [1, N]\}$ where $N$ is the number of experiments. The diversity produced by an exploration is then measured from the set of behavior space points contained in the history.

One of the key principles of an IMGEP is the reuse of past experiments. In the case of diversity search, the value of reusing experiments is illustrated by the fact that parameters leading to a specific point in behavior space can be identified more quickly when points close to it have already been found.

To integrate this principle, the process is divided in 4 steps: 1) Sampling a goal, according to a distribution parameterized by the history and the behavior space, $g_k \sim G(\mathcal{H}, \mathcal{B})$; 2) Inferring a parameter to try to reach this goal, inference is performed on the basis of a parameter sampling policy $\theta_k \sim \Pi(g_k, \mathcal{H}, \mathcal{B})$; 3) Executing a rollout of the system $R_{0\to T}(S, \theta_k)$ and getting the observation $o_k$; 4) Getting the behavior $b_k$ of the observation $o_k$, and saving the experiment outcomes in the history: $\mathcal{H} \leftarrow \mathcal{H} \cup \{(\theta_k, o_k, b_k)\}$. We refer the interested reader to [6,22] for specific examples of $G$ and $\Pi$.

## 5 Our Method

Recall that our aim is to achieve interactive exploration of a complex system guided by user's feedback without any prior knowledge, while achieving maximal constrained diversity (see Fig. 2). We implement our design choices through a variation of an IMGEP. In accordance with these choices, we constitute the behavior space $\mathcal{B}$ with a set of non-system specific features (C3). Such representation space irremediably biases the diversity produced by the process, as shown in [22]. In the following, we describe how we countermeasure this problem.

### 5.1 Augmented History

To capture the user's ROI and guide the exploration towards a specific subspace, we augment the history with classification values that emphasize the validity of the experiment in relation to the user's ROI. The history in our IMGEP variant is thus defined as $\mathcal{H} := \{(\theta_i, o_i, b_i, c_i), \forall i \in [0, N]\}$, where $c_i \in \{-1, 1\}$ corresponds to an outlier or inlier classification. Those classifications are made based on constraints formulated explicitly by the user (C1), which are constructed from features extracted from the observation space. The selected features are chosen to be intuitive and easily understandable by the user. For instance, one may use coarseness, directionality or contrast of Tamura's image features [24].

The user's ROI may thus now be seen as a simple region in a constraint space $\mathcal{C}$. As illustrated in Fig. 2-right, $\mathcal{C}$ is usually different from the behavior space $\mathcal{B}$. Indeed, constraints are only used to classify samples as inliers or outliers, they are not involved in any other calculation during the exploration process. As a result the boundaries of the ROI in behavior space $\mathcal{B}$ are not explicitly available, but this is not an issue for the IMGEP variant presented in the next section. In addition, whenever the user makes changes to the ROI during the exploration process, the exploration process does not need to be restarted since we only need to modify the classification values $c_i$ of all entries in the history.

### 5.2 Balanced IMGEP Policy

Now that we have incorporated classification values to the history, it is possible to balance the parameter selection policy $\Pi$ towards the ROI. To do so, we first sample a goal $g \sim G(\mathcal{H}, \mathcal{B})$, where $G$ is a uniform distribution of the hyper-rectangle including all behavior points found during experiments, i.e., all $b_i$ in $\mathcal{H}$ regardless of their classification. The basic nearest neighbor policy $\Pi$ works in two steps. It first identifies the parameter $\theta_k$ that leads to the behavior $b_k$ closest to the goal $g$ in the history $\mathcal{H}$, with $b_k = \arg\min_{b_i} \|g - b_i\|$. It then applies a mutation to it (for more details, see [6,22]). This policy needs to be adjusted in our case in order to maximize constrained diversity. We first want to take explicit constraints into account (C1) while maintaining a global diversity (C2), as the ROI might be subject to change during exploration. Our solution consists of introducing a balance between global diversity samples and constrained diversity samples using a meta-parameter, which is used as the probability of a Bernouilli dristibution (we use a probability of 0.5 in practice). This distribution is sampled at each experiment to determine the subset of behaviors used to determine the nearest point: in case of a global diversity sample, all entries in the history are used regardless of their classification value; in case of a constrained diversity sample, only entries corresponding to inliers (i.e., with $c_i = 1$) are taken into account.

Another difference in our case is that the behavior space $\mathcal{B}$ is not specific to the explored system (C3), which raises two issues. First, the dimensions of $\mathcal{B}$ might not be commensurate. We thus standardize both the history $\mathcal{H}$ and goal $g$ for each dimension independently, at each exploration step. Second, $\mathcal{B}$ might be high-dimensional, and our approach be subject to the curse of dimensionality: the nearest neighbor search might then pick neighbors mostly in a small subset of dimensions, hindering diversity. We mitigate that issue by using a subset of dimensions in $\mathcal{B}$, which are sampled uniformly. The number of sub-dimensions is a meta-parameter of our process (we use 3 sub-dimensions in practice).

## 6 Results

In this section, we present results obtained with the implementation of our IMGEP variant. For this purpose, two complex systems are studied. First, Gray-Scott, a reaction-diffusion system that comprises a search space with a small sub-region producing patterns; in such search spaces, random sampling is largely inefficient. Second, Lenia, a class of continuous cellular automata, in which the parameter space is high-dimensional.

### 6.1 Methodology

For each studied system, 4 exploration methods are tested:

- **Random (R)**: a random sampling of the parameter space.
- **Nearest (N)**: IMGEP exploration method, with nearest neighbors computed on the whole behavior space.
- **NearestRandomAxes (NRA)**: IMGEP exploration same as N, except that distances are computed in a 3-dimensional subspace of $\mathcal{B}$.
- **NearestRandomAxesBalanced (NRAB)**: our IMGEP variant with the full balanced policy of Sect. 5.2.

All three IMGEP methods have their $N_{init}$ first samples chosen with the Random method. The mutation applied to the candidate parameters follows a normal distribution centered at 0, with a standard deviation $\sigma_m$ given in the caption of Fig. 3. The 10-dimensional behavior space $\mathcal{B}$ is constructed from Hu moment invariants [12], the mean value of pixels, and the volume (i.e. the proportion of pixels with a value greater than a given threshold $\epsilon = 1e - 5$). The ROI of the user is explicitly given by a constraint on the volume of the observation. To satisfy this constraint, the volume must be in the range [0.6, 0.7].

### 6.2 Gray-Scott Reaction-Diffusion

Reaction-Diffusion are complex systems involving different chemical concentrations. Each concentration has the ability to diffuse into the medium and react with other concentrations or itself. Such systems were initially introduced by Turing [25], and are prone to the emergence of so-called Turing patterns (see Fig. 1a). The Gray-Scott reaction-diffusion [8] is composed of two concentrations $u$ and $v$. The kinematic of the reaction is given by the following equations:

$$\begin{aligned} \frac{\partial u}{\partial t} &= D_u \nabla^2 u - uv^2 + f(1-u) \\ \frac{\partial v}{\partial t} &= D_v \nabla^2 v + uv^2 - (f-k)v \end{aligned} \tag{1}$$

In the experiment, the update rule $U$ is the forward Euler integration of the Eq. 1 on a discrete toroidal grid of resolution $32 \times 32$. Diffusion terms are fixed to

$D_u = 0.5$ and $D_v = 0.25$. The parameter space is then composed of the feeding-rate $f$ (the quantity of $u$ constantly added) and the killing-rate $k$ (the quantity of $v$ constantly removed). The initial concentration of $u$ is set to 1 and a Perlin noise [20] thresholded at 0.7 gives the concentration of $v$.

Due to the nature of the equations, the pattern emerging in the $u$ channel is almost similar to that emerging in the $v$ channel. Therefore, the observation is constructed solely from the values of the $u$ channel.

## 6.3 Lenia

Lenia is a class of continuous cellular automaton [2,3] mainly studied for its ability to produce spatially-localized patterns. With non-locally distributed initial conditions, the system tends to produce Turing like patterns, as illustrated in the Fig. 1b. The update rule of the automata is the following:

$$A_{t+1} = A_t + dt \sum_k h_k \cdot G_k(K_k * A_t) \tag{2}$$

The growth function $G_k$ is a gaussian parametrised by $\mu_k$ and $\sigma_k$, $G_k(x) = 2\exp\left(-\frac{(x-\mu_k)^2}{2\sigma_k^2}\right) - 1$. Each kernel is made of $n$ overlapping gaussian bumps. Our implementation is based on a differentiable version of Lenia [10] which defines kernels as $K_k(x) = \sum_i^n b_i \exp\left(-\frac{\left(\frac{x}{r_k R} - a_i\right)^2}{2w_i^2}\right)$.

We simulate a Lenia rollout on a discrete toroidal grid of resolution $64 \times 64$. The system can be set up with several channels, but in our case we use only one. The number of kernels is fixed to $k = 3$, each of them having $n = 3$ bumps. Therefore the number of parameters is: $3(3 \times 3 + 4) + 2 = 41$. The initial state is set by a Perlin noise [20] (not thresholded in this case).

## 6.4 Evaluation

We rely on acceptance and diversity metrics to compare our exploration approach (NRAB) to the three alternatives methods (R, N and NRA). When randomly sampling the parameter space, 90.8% of Gray-Scott samples lead to a homogeneous pattern (i.e., a homogeneous distribution of the same pixel value), while this only occurs for 19.0% of samples with Lenia. Consequently, as both systems use the same number of initial samples ($N_{init} = 250$), the history with which IMGEP methods are initialized is more diverse for Lenia than for Gray-Scott.

We next measure the proportion of samples leading to inlier classifications of patterns, which we call the acceptance rate. The higher the rate, the better the method is able to produce samples belonging to the ROI. Results are given in Table 1 for both studied systems and all four exploration methods, without taking the initial random samples into account. Despite the low initial diversity observed with Gray-Scott, our method (NRAB) achieves an acceptance rate 15× to 20× higher than that of R and approximately 2.4× higher than N and NRA.

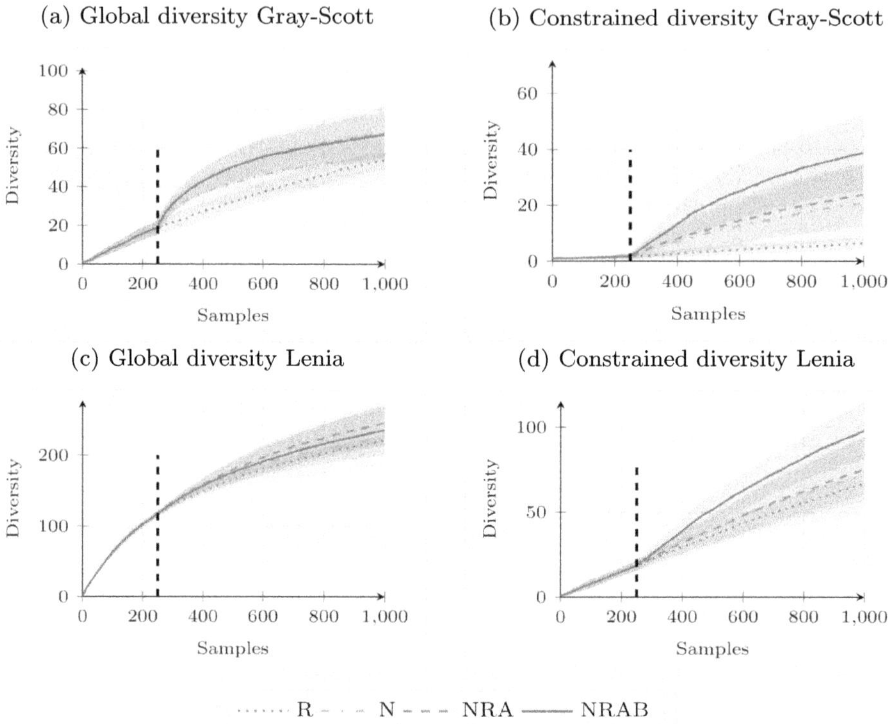


**Fig. 3. Diversity measured during exploration of Gray Scott reaction-diffusion and Lenia.** Per each configuration, a total of 50 run as been conducted with different random seed, the mean is depicted by curves, shaded areas represent std. The representation space used to measure diversity corresponds to the 13 Haralick features [11] extracted from the observation, standardized and projected in a 4 dimensional space by a principal component analysis. The $N_{init} = 250$ first samples are chosen randomly. (a) Global diversity of Gray-Scott using $N_{bin} := 200,000$. (b) Constrained diversity of Gray-Scott using $N_{bin} := 100,000$. (c) Global diversity of Lenia using $N_{bin} := 200,000$. (d) Constrained diversity of Lenia using $N_{bin} := 100,000$. We use the following parameters. For (a,b): $f \in [0.001, 0.2]$, $\sigma_m(f) = 0.2$; $k \in [0.01, 0.075]$, $\sigma_m(k) = 0.001$. For (c,d): $R \in [2, 40]$; $T = \frac{1}{dt} \in [2, 20]$; $\mu_k \in [0.05, 0.5]$; $\sigma_k \in [0.001, 0.18]$; $h_k \in [0.01, 1]$; $r_k \in [0.2, 1]$; $b_i \in [0.001, 1]$; $w_i \in [0.01, 0.5]$; $a_i \in [0, 1]$. All $\sigma_m$ are equal to 0.2, except $\sigma_m(T) = 0.5$ and $\sigma_m(\sigma_k) = 0.01$.

The acceptance rate of NRAB on Lenia remains the highest, even though less so (2× to 2.3× higher than R, 1.6× higher than N or NRA). In both systems, given the time required to evaluate a sample, the NRAB method finds samples belonging to the ROI more quickly.

To measure global diversity, we choose a metric already used in quality-diversity search [21]. It works by partitioning the representation space into $N$ bins. Each observation is embedded in the associated bin of the behavior space.

**Table 1.** Acceptance rates. For each exploration method: the percentage of valid samples, the time required to evaluate a sample and the time required to obtain a valid sample are given (Measurements are taken with Intel Core i7-4790K CPU and Nvidia Titan V GPU).

| | Gray-Scott | | | | Lenia | | | |
|---|---|---|---|---|---|---|---|---|
| | R | c\|N | NRA | NRAB | R | N | NRA | NRAB |
| Acceptance rate | 0.68% | 5.35% | 5.6% | **13.66%** | 8.33% | 11.7% | 10.99% | **19.46%** |
| Time/sample (ms) | **155** | 353 | 388 | 412 | **767** | 796 | 823 | 845 |
| Time/inlier (s) | 22.784 | 6.598 | 6.928 | **3.016** | 9.207 | 6.803 | 7.488 | **4.342** |

Diversity then corresponds to the number of bins in which at least one observation is embedded. The constrained diversity metrics is computed the same way, except that we *only consider inliers*. Results are shown in Fig. 3 for both types of diversity and the two studied systems. A qualitative assessment of these results is shown in Fig. 1.

As expected, IMGEP methods perform better than random parameter sampling. NRA leads to greater global diversity than N, which is expected given the variance in the distribution of behaviors along the 10 axes. The global diversity of NRAB is very close to that of NRA, thus balanced sampling allows for global diversity to be maintained. IMGEP methods appear to be less effective on Lenia. As mentioned above, the rate of homogeneous patterns produced by uniform sampling of Lenia's parameter space is low. Furthermore, compared with Gray-Scott, Lenia offers greater variation in the patterns it produces. Thus, for a few samples, random sampling already produces a significant diversity. The effectiveness of the IMGEP method for Lenia is more marked when a larger number of samples is employed: as a comparison, Reinke and al. and Etcheverry and al. [6, 22] used 1,000 initialization samples for a total of 5,000 samples.

Most importantly, our method achieves greater constrained diversity than the others. The effectiveness of balanced sampling is demonstrated by the increase in constrained diversity of NRAB compared to NRA, for both Gray-Scott and Lenia. We further show in Fig. 4 the impact of the balancing meta-parameter on constrained diversity. A value of 1 (only sampling the ROI) yields high constrained diversity but low global diversity; whereas a value of 0 (sampling the whole space) yields the opposite situation, which is prone to local minima issues. Our choice of a rate of 0.5 thus appears as a safe compromise.

## 7 Discussion

In this paper, we have considered ways to incorporate interactive human guidance in complex system exploration. In practice, we employ a modified IMGEP, augmenting the history to accommodate for variable user ROI. We use a balanced policy to sample the parameter space, which maintains global diversity while achieving better constrained diversity and acceptance rates. As shown in

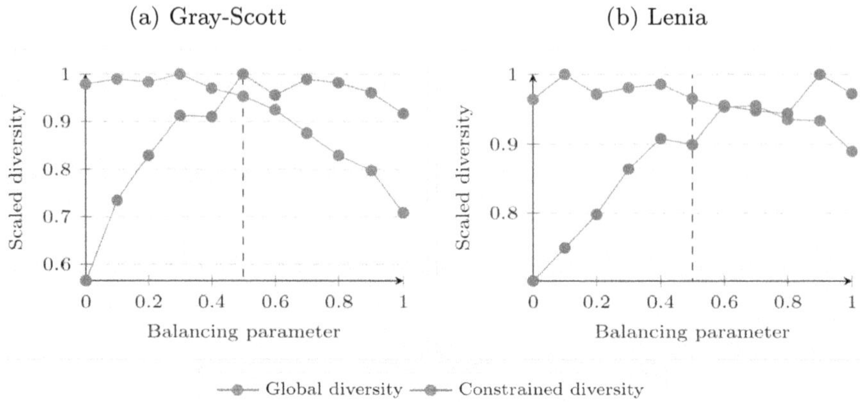


**Fig. 4. Study of the balancing meta-parameter.** The conditions are similar to those described in the Sect. 6. Each point represents the diversity after 1000 samples using NRAB method, divided by the maximum diversity encountered, for (a) Gray-Scott and (b) Lenia.

our experiments on Gray-Scott and Lenia reaction-diffusion, our method achieves better results while remaining agnostic to the mapping between system parameters and behaviors. We now discuss each of our design choices in turn.

The use of explicit constraints (C1) allows for the ROI to be changed without losing information from past samples, as all samples lead to a greater global diversity. However, not all regions of interest can be formulated through intuitive explicit constraints. To get around this problem, we might consider using implicit constraints (i.e., labelling example inliers/outliers from exploration samples).

As shown in Fig. 4, maintaining global diversity (C2) is beneficial. However, the choice of balance meta-parameter depends on the user ROI and considered system. A dynamic balancing based on statistical measurements or user feedback could be explored in future work. It is also possible to use a phase space approach to adapt the parameter sampling policy. As shown by Papadopoulos [19], phase transition zones tend to bring greater diversity, which is beneficial for global diversity but needs to be reconsidered in the case of constrained diversity.

The definition of a non-specific behavior space (C3) eliminates the need for prior knowledge of the system outputs. However, the exploration process could be augmented with priors without altering the overall behavior. When available, apriori knowledge of the system behavior could lead to the construction of a more effective behavior space that embeds a greater diversity (e.g., more initialization samples to get around Gray-Scott's low acceptance rate).

In future work, we would like to consider two extensions of our approach. We chose not to incorporate the initial state in the search space, in order to only focus on the dynamic behavior of the system as a function of the update rule. However, the methods employed could be used to explore all the basins of attraction in behavior space of a complex system for a fixed update rule. Furthermore, this work is focused on systems producing image/texture output, but it is possible to extend the scope of the considered systems to those producing any other type of output. In particular, those whose observation corresponds to system dynamics (i.e., $O(R_{0 \to T}(\mathcal{S}, \theta_i)) := R_{0 \to T}(\mathcal{S}, \theta_i)$), such as swarm/flock/crowd models [23].

## References


1. Ball, P.: The Self-Made Tapestry: Pattern Formation in Nature. Oxford University Press (1998). https://doi.org/10.1093/oso/9780198502449.002.0001
2. Chan, B.W.C.: Lenia: biology of artificial life. Complex Syst. **28**(3), 251–286 (2019). https://doi.org/10.25088/ComplexSystems.28.3.251
3. Chan, B.W.C.: Lenia and expanded universe. In: The 2020 Conference on Artificial Life, pp. 221–229. MIT Press, Online (2020). https://doi.org/10.1162/isal_a_00297
4. Dorin, A., Stepney, S.: What is artificial life today, and where should it go? Artif. Life **30**(1), 1–15 (2024). https://doi.org/10.1162/artl_e_00435
5. Etcheverry, M.: Curiosity-Driven AI for ScienceâĂŕ: Automated Discovery of Self-Organized Structures. Ph.D. thesis, Université de Bordeaux (2023)
6. Etcheverry, M., Moulin-Frier, C., Oudeyer, P.Y.: Hierarchically organized latent modules for exploratory search in morphogenetic systems. In: Proceedings of the 34th International Conference on Neural Information Processing Systems, NIPS 2020, pp. 4846–4859. Curran Associates Inc. (2020)
7. Forestier, S., Portelas, R., Mollard, Y., Oudeyer, P.Y.: Intrinsically Motivated Goal Exploration Processes with Automatic Curriculum Learning (2017). https://doi.org/10.48550/arXiv.1708.02190
8. Gray, P., Scott, S.K.: Autocatalytic reactions in the isothermal, continuous stirred tank reactor: oscillations and instabilities in the system A + 2B → 3B; B → C. Chem. Eng. Sci. **39**(6), 1087–1097 (1984). https://doi.org/10.1016/0009-2509(84)87017-7
9. Grillotti, L., Cully, A.: Unsupervised behavior discovery with quality-diversity optimization. IEEE Trans. Evol. Comput. **26**(6), 1539–1552 (2022). https://doi.org/10.1109/TEVC.2022.3159855
10. Hamon, G., Etcheverry, M., Chan, B.W.C., Moulin-Frier, C., Oudeyer, P.Y.: Discovering Sensorimotor Agency in Cellular Automata using Diversity Search (2024). https://doi.org/10.48550/arXiv.2402.10236
11. Haralick, R.M., Shanmugam, K., Dinstein, I.: Textural features for image classification. IEEE Trans. Syst. Man Cybern. **SMC-3**(6), 610–621 (1973). https://doi.org/10.1109/TSMC.1973.4309314
12. Hu, M.K.: Visual pattern recognition by moment invariants. IRE Trans. Inf. Theory **8**(2), 179–187 (1962). https://doi.org/10.1109/TIT.1962.1057692
13. Kingma, D.P., Welling, M.: Auto-Encoding Variational Bayes (2022). https://doi.org/10.48550/arXiv.1312.6114
14. Kumar, A., et al.: Automating the Search for Artificial Life with Foundation Models (2024). https://doi.org/10.48550/arXiv.2412.17799
15. Ladyman, J., Lambert, J., Wiesner, K.: What is a complex system? Eur. J. Philos. Sci. **3**(1), 33–67 (2013). https://doi.org/10.1007/s13194-012-0056-8
16. Mitchell, M., Crutchfield, J.P., Das, R.: Evolving cellular automata with genetic algorithms: a review of recent work. In: Proceedings of the First International Conference on Evolutionary Computation and Its Applications (EvCA 1996), vol. 8 (1996)

17. Mordvintsev, A., Randazzo, E., Niklasson, E., Levin, M.: Growing neural cellular automata. Distill **5**(2), e23 (2020). https://doi.org/10.23915/distill.00023
18. Mouret, J.B., Clune, J.: Illuminating search spaces by mapping elites (2015). https://doi.org/10.48550/arXiv.1504.04909
19. Papadopoulos, V., Doat, G., Renard, A., Hongler, C.: Looking for complexity at phase boundaries in continuous cellular automata. In: Proceedings of the Genetic and Evolutionary Computation Conference Companion, GECCO 2024 Companion, pp. 179–182. Association for Computing Machinery (2024). https://doi.org/10.1145/3638530.3654420
20. Perlin, K.: An image synthesizer. ACM SIGGRAPH Comput. Graph. **19**(3), 287–296 (1985). https://doi.org/10.1145/325165.325247
21. Pugh, J.K., Soros, L.B., Szerlip, P.A., Stanley, K.O.: Confronting the challenge of quality diversity. In: Proceedings of the 2015 Annual Conference on Genetic and Evolutionary Computation, pp. 967–974. ACM (2015). https://doi.org/10.1145/2739480.2754664
22. Reinke, C., Etcheverry, M., Oudeyer, P.Y.: Intrinsically motivated discovery of diverse patterns in self-organizing systems. In: International Conference on Learning Representations (2020). https://doi.org/10.48550/arXiv.1908.06663
23. Reynolds, C.W.: Flocks, herds and schools: a distributed behavioral model. In: Proceedings of the 14th Annual Conference on Computer Graphics and Interactive Techniques, SIGGRAPH 1987, pp. 25–34. Association for Computing Machinery, New York, NY, USA (1987). https://doi.org/10.1145/37401.37406
24. Tamura, H., Mori, S., Yamawaki, T.: Textural features corresponding to visual perception. IEEE Trans. Syst. Man Cybern. **8**(6), 460–473 (1978). https://doi.org/10.1109/TSMC.1978.4309999
25. Turing, A.M.: The chemical basis of morphogenesis. Philos. Trans. Roy. Soc. London Ser. B Biol. Sci. **237**(641), 37–72 (1952). https://doi.org/10.1098/rstb.1952.0012
26. Zhang, J., Lehman, J., Stanley, K., Clune, J.: OMNI: Open-endedness via Models of human Notions of Interestingness (2024). https://doi.org/10.48550/arXiv.2306.01711

# A Phenomenological Glider? Enactivism and IIT in the Game of Life

Fernando Rodriguez(✉) and Phil Husbands

AI Research Group & Sussex Centre for Consciousness Science, University of Sussex, Brighton, UK
{f.rodriguez-vergara,p.husbands}@sussex.ac.uk

**Abstract.** In this paper we examine the potential relation between autonomy and phenomenological properties as purported by the Integrated Information Theory of Consciousness (IIT). Our central motivation in this case is that, given the strong intuition that a phenomenological dimension requires a living organism to support it, and the opposite claim from IIT—that this can be explained as a consequence of a particular type of system dynamics, without the need of high complexity, less even of life—, we deemed it interesting to see if a minimal autonomous system could meet the IIT criteria for integrating information. For this we measured intrinsic information from a glider in the Game of Life (GoL) cellular automata, a structure that has been purported to be an autonomous/autopoietic pattern in the GoL. Results seem to indicate a relation between intrinsic information and cognition in enactive terms.

**Keywords:** Phenomenology · Enactivism · IIT · Artificial Life · GoL

## 1 Introduction

Roughly, enactivism is the view that cognition arises through a dynamic interaction between an acting organism and its environment. One of its foundational axes [54] has been its emphasis on the necessary relation between life, cognition, and phenomenological experience [16,21], which would ultimately be rooted in the organizationally recursive nature of autonomous systems and, specifically, in the autonomy of life as a special kind [18,48,56,59]. However, in spite of advances, there is still no conclusive understanding about the nature of this relation; a conundrum that has led to splintering enactivist positions.

In this paper, we present an attempt to better develop some of these ideas more formally, by exploring the feasibility of introducing principled information measurements based on the mathematical framework proposed by the Integrated Information Theory of Consciousness (IIT) [4,37], using the Game of Life (GoL) cellular automata [13,22] as ground for illustration and analysis.

The measure of information that we used, cause-effect information, is based on the metrics proposed by the IIT [37] (see [50] or [38] for an introduction),


M. Bianchini et al. (Eds.): WIVACE 2025, CCIS 3001, pp. 129–143, 2027.
https://doi.org/10.1007/978-3-032-33185-4_10

first presented as such in [49] and currently refined into its fourth version [4]. We centered our attention around the notion of intrinsic information, because in spite of the strong assumptions posited by IIT, it has a relatively direct application, which we believe, can be useful to sharpen the conceptual and mathematical description of a subjective/intrinsic point of view of autonomous systems in terms of information. In fact, the notion of an intrinsic perspective is an essential agreed enactive premise, which hasn't been fully unraveled formally.

In this regard, the use of toy models can be very helpful, as they enable more exhaustive descriptions and analyses, often intractable in more complex models [9,24]. Random boolean networks and cellular automata are popular examples of such models [8,10,11,23,27,43,53,62].

Considering that: *i*) autonomy is an underlying property of cognitive systems and, by extension, of the more concrete examples we can consider to be conscious systems. And *ii*) that as postulated by IIT, the capacity for integrating information seems to be deeply related to conscious experience; we have decided to investigate the relation between these two properties, to see whether the formulations of the IIT can be applied to a minimal case of autonomy.

In brief, we present an analytical exploration of the compatibility between IIT's and enactive premises. If there is some relation to be found, it may be a hint as to the reason for the differences we believe to exist between organic and inorganic possibilities for consciousness. If not, it may suggest a deeper division between autonomy and experience than we had previously thought to exist, and act as a call to reevaluate some of our central assumptions in cognitive science.

## 2 Understanding the Problem

Stemming from the notion of autopoiesis [30], but in order to account for a more general notion of *autonomy* beyond the cellular case, Varela [52] introduced the formal concept of *organizational closure*, characterizing autonomous behavior in terms of operations within a self-referential space of transformations. Autonomy, in this context, would be the product of a logic intrinsic to the system, therefore a fundamental kind of (subjective) intelligence, which would be gradual and incremental.

Building on this notion, in *The Embodied Mind* [54] autonomy is further characterized as structurally embodied and environmentally embedded, giving rise to the concepts of *enaction* and enactive autonomy. These concepts are clearly illustrated by means of Bittorio, a minimally autonomous cellular automata described as capable of *bringing forth a domain of significance* by selectively enacting external regularities (hence, displaying a minimal form of cognitive behavior), but considered obviously devoid of experience (in contrast to color perception in humans) [54,55].

Beyond the specifics of the experiment, the fundamental point is the gap between these two systems, namely, between the raw cognitive behavior displayed by Bittorio and the phenomenological experience associated with it in the human case; why/how, on what grounds, do some varieties of autonomy entail sentience and not others?

Enactivism claims that insofar as the autonomy of living systems is the process of the constitution of a precarious identity, it would account for the origin of cognition, while providing every organism with a referential perspective for interactions, intentionality and semantic meaning in phenomenological terms [16,28,47,56]. Put differently, the need to act for subsistence would give rise to a relevance of external entities and events with respect to its intentional states; the enactive *sense-making*.

Phenomenological experience, as for us, would be the result of this biological underpinning trait (i.e. sense-making), plus the particular nested nature of the autonomy of the nervous system. In other words, while life would entail cognition, thus a specific kind of *blind* biological autonomy, nervous systems would entail consciousness; a specific kind of *phenomenological* autonomy.

Nevertheless, in spite of enactivist research, there is still no conclusive, principled explanation for how the specific autonomy of life would entail the appearance of either of these, an intentional mind, or a phenomenological observer [1,19,25,46]. As a matter of fact, this is one of the most intriguing issues in current cognitive science and, understandably, has led to variegated positions:

Roughly speaking, bioenactivism still defends the original life-mind continuity thesis [17,45], radical enactivism argues for a *mechanical* view of simple organisms [25,26], a resurgence of the autopoietic theory of cognition proposes that consciousness is exclusive of higher order systems at least capable of language [32,57], while other approaches have returned towards rather representational explanatory stances [15,44,61]. See [7,20,41,58], for some good overviews about the state of affairs.

To put it simply, on the one hand, it seems evident that from the arbitrary set of entities to which we could attribute consciousness, beside us, the more likely case is that of other living beings (a dog or a whale, for example). This, of course, is only an intuition, but a very strong one nonetheless. On the other hand, we have no real proof whatsoever; Why living beings? Actually, when we think of organisms such as sunflowers, any single cell composing our organs, bacteria, or amoebae, to name a few, we suddenly realize that attributing sentience to every form of life may be precipitated.

## 3 Intrinsic and Integrated Information

In an attempt to reconcile ideas, we have invoked the notion of integrated information, as presented in the IIT framework [4,5,37,49], which may provide a link in this sense. While we do not fully commit to IIT premises, the idea that integrated information may be somehow implicated in the existence of consciousness is appealing, as it could provide us with an operational criteria.

As a fundamental axiomatic principle, IIT portrays the *existence* of experience as immediate and irrefutable [4] (in a similar spirit to Descartes' *I think therefore I am*). Along the lines of Maturana [31,32], the point being made here is that physical phenomena is somehow subordinated to our ontologically prior subjective observation.

Thus, given that we can only be sure of our experience, if we believe that there may be things that exist independently of our observation, then the *signature* of their physical existence would be their operational reliability (logical cause and effect) under the manipulation of a conscious –therefore real– observer [4]. In other words, the existence of physical entities and their behavior is something that has to be operationally assessed from within our conscious experience, and the proposed criterion to assess them is causality, specifically cause-effect power.

Central axiomatic properties of conscious systems in IIT, which can be determined through the calculation of the cause-effect power of a system, are intrinsic information and integration. The former is to be understood as the reduction of uncertainty (in the classical sense), but from the perspective of the system itself, so that the information generated by the system is proportional to the number of alternative potentially distinguishable scenarios that are ruled out whenever the system makes some distinction (specifies a state). The latter aims to account for the unified and irreducible nature of experience, and reflects the extent to which information generated by the system as a whole is more than that of its composing parts taken independently [4,29,37,38].

For clarity and simplicity's sake, we will make use of the toy model example presented in [37] (see Fig. 1), where, given a system of mechanisms A, B and C, cause (Eq. 1), effect (Eq. 2) and cause-effect (Eq. 3) information can be obtained from:

$$ci = D(p(ABC^p \mid A^c = 1) \mid\mid p^{uc}(ABC^p)) \tag{1}$$

$$ei = D(p(ABC^f \mid A^c = 1) \mid\mid p^{uc}(ABC^f)) \tag{2}$$

$$cei = min(ci, ei) \tag{3}$$

Cause-effect information ($cei$, Eq. 3) is the measure of how relevant the effect of the structural selectivity of the system is to the system, from the *perspective of the system itself*, which is given by the shared causal power measured from cause ($ci$) and effect ($ei$) information.

$p(ABC^p \mid A^c = 1)$ and $p(ABC^f \mid A^c = 1)$, called cause and effect *repertoires*, respectively, are the constrained probability distributions for past and future system's states, given that the state of $A = 1$. The superscripts $^p$,$^c$ and $^f$ stand for past, current and future, and $^{uc}$ denotes unconstrained distributions; whereas D represents the distance between distributions, measured using the earth mover's distance (EMD) method [42,60].

Cause information (Eq. 1) is obtained by comparing the *cause repertoire* with the unconstrained distribution of the previous state. In simple terms, cause information would reflect the fact that the state transition obeyed a causal structure instead of chance, and provides a measure of how powerful (how informative) that transition was by excluding other alternatives.

Similarly, effect information is obtained by taking the distance between the causally constrained *effect repertoire* and the unconstrained distributions of

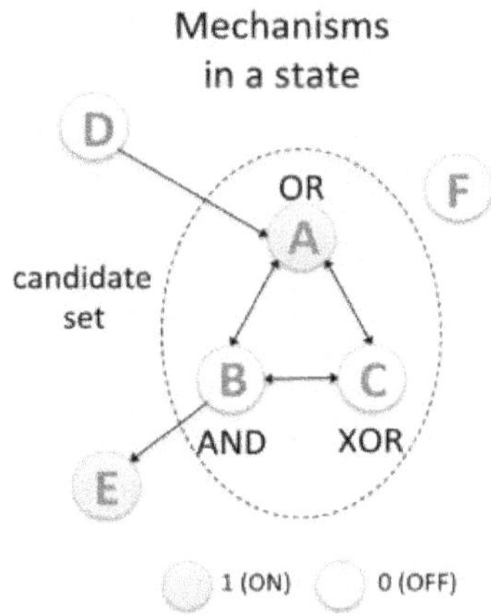


**Fig. 1.** Toy model example from version (3.0) of IIT. In order to evaluate whether the elements A, B and C conform together a system ABC, the amount of cause-effect information is calculated for every subset of the powerset. The system is defined by the subset with the higher integrated information.

future states, measuring how informative the mechanism is, with respect to others with no effect power. However, whereas the unconstrained past is formalized as a simple homogeneous distribution (i.e., understood as causally inefficient and/or uninformative), the unconstrained future distribution is taken as independent of the current state of the system, although still dependent on its causal structure—basically, as if the system did not have any causal influence upon itself, hence, as if it was fully externally determined.

Cause-effect information (Eq. 3) is the minimum between them because cause-effect mechanisms provide information, or have cause-effect power, to the extent that their structure excludes alternative past-to-future possibilities.

Since integration is intended to account for the conscious experience of *one* system, made-up from the integration of otherwise independent elements, the cause-effect information generated by the system must be irreducible to a composition of the cause-effect information produced by its mechanisms taken independently. This intrinsic irreducibility is quantified as integrated information (denoted as $\phi$), by partitioning the cause-effect structure of the system according to its weakest link and using the partitioned system to look for an excess of information. Thus, if the cause-effect information provided by them is the same, this is because the former system can be reduced to the later.

The procedure is quite straightforward (albeit computationally incredibly expensive), bearing in mind how we obtained the cause-effect information: The cause repertoire of the system is compared with the cause repertoire of the partitioned system; then the effect repertoire of the system is compared with the effect repertoire of the partitioned system; finally, integrated information is obtained by taking the minimum between the excess of cause and effect exceeding integrated information.

## 4 Gliders in the Game of Life

Conway's Game of Life is a cellular automaton and a zero-player game, in which the player can only choose the initial state of a grid of cells that can be *alive* or *dead* (represented by symbols like 1 and 0, or ON and OFF). After the setup, the grid is automatically updated in discrete timesteps, by continuously applying the same rule to each of the cells on the grid [13,22].

The rules are the following: 1) Reproduction: every *dead* cell surrounded by 3 active cells, comes to life. 2) Overcrowding: every active cell surrounded by more than 3 active cells *dies*. 3) Loneliness: Any *living* cell surrounded by less than 2 active cells *dies*. Sometimes rules 2) and 3) are also expressed as: *living* cells surrounded by 2 or 3 active cells remain *alive*, otherwise they *die*. For an illustration of different initial states and the updating rules, see Fig. 2.

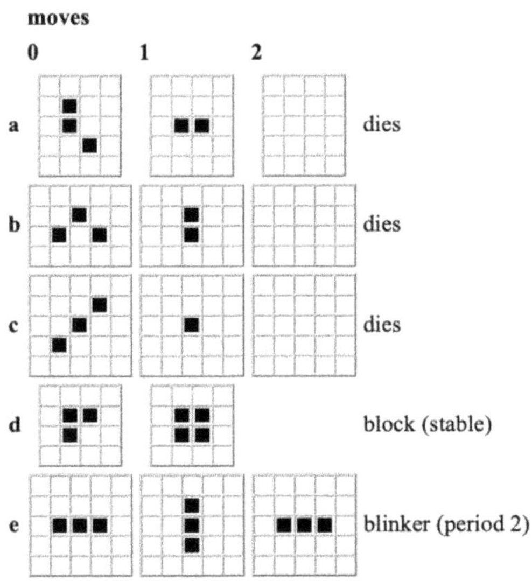


**Fig. 2.** Example of GoL grids evolution in time. Different initial states lead to different state transitions by the update of every cell on the grid at the same time. See text for further details. (Image taken from [22]).

In the Game of Life, a **glider** is an autonomous dynamical pattern composed of 5 active cells, which, in an empty environment will loop indefinitely through 4 states (this looping behavior is illustrated in Fig. 3).

Beer [11,12] first proposed a glider as a model of an autopoietic system. The intuition behind this association is that a glider basically operates as a coherent self-producing network of 16 structures moving across the lattice of the GoL, subordinated to a spatiotemporal organization (Fig. 4).

An important difference that this case introduces, when contrasted with the ABC system (Fig. 1), is that, for every state transition, a glider production

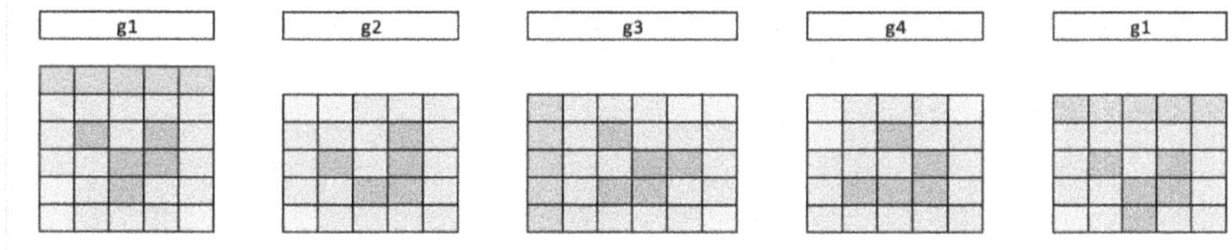


**Fig. 3.** Example of the state cycle of a glider in the Game of Life. Green cells in the image represent active cells in the lattice. In order to recognize or to associate a given pattern to be a glider, the cells surrounding those 5 active cells must be deactivated, this is represented by the blue cells. Pink cells, represent cells that may be active or not. Lastly, gray cells represent cells that were occupied by the glider (either by active or non-active cells) in the previous timestep. (Color figure online)

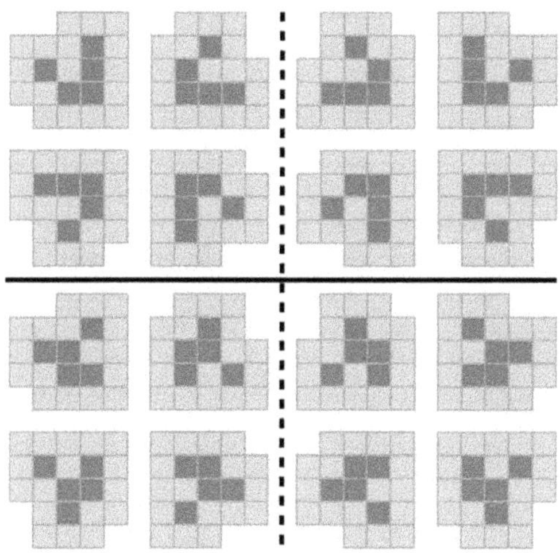

**Fig. 4.** Depiction of the 16 configurations of a glider in the Game of Life. Following [11], these are taken to be the structurally valid instances of the operationally closed domain of the Glider. Original image and further details can be found in [11].

process specifies 2 destruction, 3 maintenance and 2 creation processes, in other words, at every timestep the glider is composed of a new set of cells (see Fig. 5). Thus, there is a constant loss of information towards the past, due to cells that are not part of the glider anymore, and to the future, because the current elements of the system cannot specify states other than their own, at least within the context of intrinsic information.

Put differently, the dynamic re-instantiation of the glider introduces a degree of (intrinsic) information loss, or underdetermination, because some past and future elements are inaccessible from the current purviews available to the elements of the candidate system. (Given that we cannot assume what would be the system by the end, these repertoires have to be calculated over different combinations of elements and the particular set of elements over which the repertoire is calculated is called its purview).

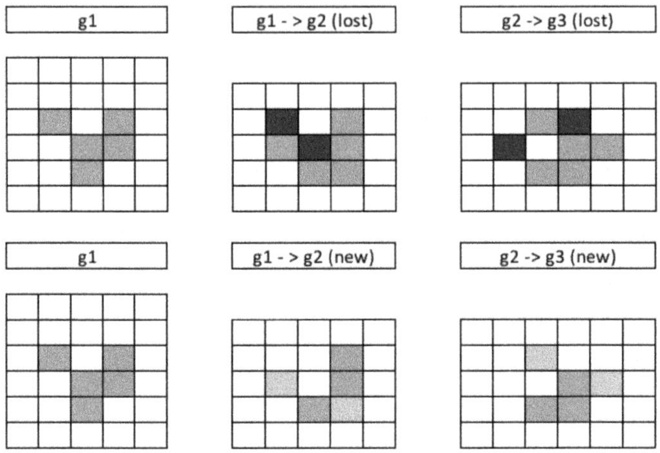


**Fig. 5.** Two sequences of images illustrating the same glider state transitions from $g1 \rightarrow g2 \rightarrow g3$. In the above sequence, blue cells represent active cells, while black cells represent cells that were active in the previous timestep, but not in the present state of the system. Conversely, in the bottom sequence, green cells represent cells that are active in the present state of the system, but that were previously non-active (i.e., *new* cells). The state space changes as a consequence of the structural transformations. (Color figure online)

In this sense, and insofar as IIT formulations are supposed to be applied to temporally discrete states (which represent the present of the system), we computed intrinsic information considering the glider made of the active cells at each timestep to be the candidate system. Hence, for every element within a time-frame, we obtained the corresponding probability distributions expressing the likelihood of them being alive (active) or not in the previous and in the subsequent timesteps. Basically, the idea is that, as emphasized in [30], the elements of the system may change, but as long as the system replace them with new elements that are operationally equivalent, the organization of the system remains the same (e.g., cells in our body are constantly replaced when dead).

Intrinsic information not only has to be calculated for every subset of the powerset made from the elements of the candidate system, but it also has to be evaluated –as if– from the perspective of each of these subsets. In other words, each subset has to be evaluated from every subset of the same powerset. More concretely, in the current case, for both past and future transitions, each of the 31 subsets representing possible systems had to be evaluated from the same 31 subsets representing the possible purviews. In order to do so, we defined valid and invalid purviews according to the same criteria we used for the possible systems (this is illustrated in Fig. 6).

A limitation we should address at this point concerns the final step of the procedure as proposed by the formulation of the IIT, namely, the calculation of integrated information. Concretely, even for our toy experiment, the calculations became impossible given temporal and computational constraints, therefore compelling us to look for an alternative way to approximate the principle of integration. In fact, the algorithm for integration through exhaustive partitioning and comparison (as proposed in [37]) has received criticism on philosophical and mathematical grounds [2,8,14,33–36,39,51]. Accordingly, taking inspiration from the discussions partially summarized in [6], we decided to follow [2,3] and proceed with the proposed computations until obtaining the $\phi$ measure (i.e., the source of the maximum cause-effect information). Then, in order to assess integration, we compared these values under the premise that synergistic information will yield higher values than the sum of all the individual components.

## 5 Results

The results we obtained can be summarized in three points: *i*) intrinsic information calculations reveal a reduction into only 2 base configurations (see Fig. 4); *ii*) the maximum source of cause-effect information is localized in smaller subsets than the system as a whole; and *iii*) these smaller subsets are synergistic in nature.

First, as noted in [11], the 16 possible configurations of the glider can be reduced to only 2 different types, under symmetry operations, namely, translation, transposition (mirroring) and rotation. While this is assessed visually in (Fig. 4), we wanted to check whether this could be also examined through information values. To this end, we computed the *phi* values for every element individually for every configuration of the glider, from the purview of the glider itself and then added them to see if the visual observation held.

Indeed, as it is shown in Table 1, the numerical results confirm the symmetrical reduction to only two configurations (we will refer to these configurations as C1 and C2 from now on). These results do not only apply to the direct summation, but are also obtained from the full calculations of the candidate mechanisms and purviews obtained by applying Eqs. 1, 2 and 3.

The second part of results concerns the maximum cause-effect information values, which are 1.84 for C1 and 1.65 for C2. Interestingly, although the values are different from the ones obtained from the direct summation of the elements presented in Table 1, the symmetry reduction to only 2 configurations holds, as well as the fact that one of these configurations (C1) is slightly more informative than the other (C2), therefore ratifying the intuitive conclusion from [11].

On the other hand, however, the source of the maximum cause-information values we obtained were not only those given by the whole system, but also by smaller subsets, entailing that the causal power of the system and therefore, along the lines of the IIT, the system itself, can be reduced to these elements. Specifically, for C1 the same amount of information that can be measured from the system as a whole is also generated by a subgroup of 4 active cells. And for C2, by a subgroup of only 3 elements.

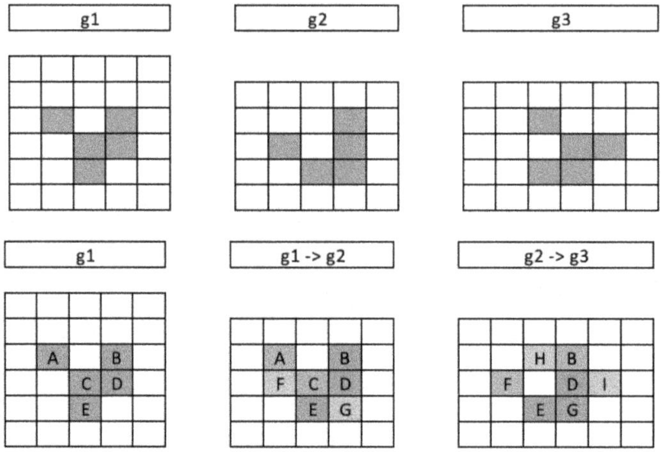


**Fig. 6.** The above and bottom images represent the same sequence of state transitions $g1 \rightarrow g2 \rightarrow g3$. The images above simply depict 3 different sequential configurations, while in the image at the bottom, the same cells are labeled by letters A-H to illustrate the valid elements and correlated purviews. Blue cells represent active cells. Gray cells represent cells that were active in the previous time-step, but not in the time-frame considered as the present of the system. Green cells represent newly active cells, i.e. cells that are active in the present state, but were not previously active. (Color figure online)

The composition of these subgroups (the specific elements yielding the maximum amount of information) depend only on the orientation of the glider, which conceptually matches the reduction to C1 and C2 on symmetry bases. The fact that for C2 the maximal subset has a smaller number of active cells may be somehow related to the lower amount of information given by the glider as a unit (or viceversa), or perhaps because to its particular structural disposition. Nevertheless, we could not determine the concrete reason only from our numerical data.

Finally, as previously mentioned, in order to determine whether information could be considered to be synergistic (hence, a consequence of integration) we compared the groups that produced the maximum cause-effect information values as a whole, against the sum obtained from their individual elements. In this case, the difference was quite high; for C1 the excess of information was 0.71, whereas for C2, only 0.03. This is most probably due to the mechanism/purview ratio, because for C1 the most informative cause-effect mechanism is composed of 4 cells, but the purview only of 3 of them. For C2, however, there are the same number of cells for both. In other words, given that the purview is limited to 3 cells (because past and future states have 2 inaccessible cells), the fact that for C1 the causal structure has one more cell seems to imply correlated higher value from the comparison of the probability distributions (i.e., entails more surprise).

**Table 1.** Table of the cause-effect power for each element and configuration of the glider, together with their resulting sum. Each 'cfg' stands for different spatial configuration of the glider, according to type, orientation and chirality (see Fig. 4 for an illustration). As it can be seen, the 16 different structural configurations can be reduced to just two types (as indicated by SUM values, and the combinations of cause-effect powers). Further details about this interpretation in main text.

| phi | A | B | C | D | E | SUM |
|---|---|---|---|---|---|---|
| cfg = 0 | 0.174 | 0.456 | 0.403 | 0.287 | 0.207 | 1.530 |
| cfg = 1 | 0.166 | 0.267 | 0.369 | 0.192 | 0.265 | 1.260 |
| cfg = 2 | 0.456 | 0.287 | 0.174 | 0.403 | 0.207 | 1.530 |
| cfg = 3 | 0.166 | 0.192 | 0.266 | 0.265 | 0.369 | 1.260 |
| cfg = 4 | 0.287 | 0.207 | 0.456 | 0.403 | 0.174 | 1.530 |
| cfg = 5 | 0.192 | 0.265 | 0.266 | 0.369 | 0.166 | 1.260 |
| cfg = 6 | 0.174 | 0.403 | 0.207 | 0.287 | 0.456 | 1.530 |
| cfg = 7 | 0.369 | 0.266 | 0.265 | 0.166 | 0.192 | 1.260 |
| cfg = 8 | 0.207 | 0.287 | 0.403 | 0.456 | 0.174 | 1.530 |
| cfg = 9 | 0.265 | 0.192 | 0.369 | 0.266 | 0.166 | 1.260 |
| cfg = 10 | 0.207 | 0.403 | 0.174 | 0.287 | 0.456 | 1.530 |
| cfg = 11 | 0.369 | 0.265 | 0.266 | 0.192 | 0.166 | 1.260 |
| cfg = 12 | 0.174 | 0.403 | 0.456 | 0.207 | 0.287 | 1.530 |
| cfg = 13 | 0.166 | 0.369 | 0.266 | 0.265 | 0.192 | 1.260 |
| cfg = 14 | 0.456 | 0.287 | 0.207 | 0.403 | 0.174 | 1.530 |
| cfg = 15 | 0.192 | 0.166 | 0.265 | 0.266 | 0.369 | 1.260 |

## 6 Conclusions and Further Work

There is yet no theory capable of bridging the cognition-sentience gap. In fact, we could say that these are the current limits of enactivism [19,41]. A radically different attempt to breach this gap has been presented by IIT [4,37], which has encouraged objective research around a generally 'ethereal' topic.

To test the compatibility between these ideas, we have applied measures of intrinsic information to a glider in the Game of Life (GoL) as a toy experiment.

To this end, we have analyzed its state transitions and measured cause-effect power as proposed by the IIT, considering GoL to be a well suited toy scenario for a minimal, initial approach, being more feasible to evaluate.

We have shown that, after adaptation, it is possible to apply equations from IIT to a system that is not materially static (i.e., the glider, as opposed to the ABC system in Fig. 1), which can provide us with a different insight about causality and closure. Moreover, we have shown that this particular information measures support enactivist ideas about minimal forms of life and cognition.

Centrally, our results support the idea that IIT is operationally assessing the existence of a system as a real unit (i.e. independent of an observer), something that enactivism has emphasized, but mostly described just conceptually. However, the step from these blind cognitive mechanisms to sentience still requires a strong axiomatic assumption—be it life or causality-as-existence. In this sense, enactivism and IIT seem to share a research strategy.

Thus, IIT formal tools may provide a solid ground for important enactivist ideas, even though the scope of application would necessarily have to change from consciousness to protocognition or intentionality, insofar as causally emergent material integration may explain ur-intentional processes at most [1,25].

Apart from the simplicity of GoL itself, two important limitations are the examination of only isolated cases (no environmental interaction), and the calculations using only the ON cells of the glider. Specifically, given computational constraints we considered the pattern formed by the 'alive' cells as the candidate system to analyze, which may be problematic from a strict IIT point of view—from which probably the best approach would have been to analyze the ON, plus the OFF cells (i.e. the active pattern, plus its membrane).

Notwithstanding, the main issue we have pointed out here remain the same, namely, the continuous creation and destruction of elements that travel through the grid together. These could be considered a global amalgamated system, as well as a non-integrated ensemble (let's say, in the sense of *Physalia physalis*).

Along these lines, even if we had to find an alternative for the final integration step, the toy experiment allowed us to measure cause-effect power for every mechanism and purview, and for each configurations of the glider. In fact, our results coincide with previous formal developments and intuitions [11,12], thus suggesting that at least in some cases, simplified versions for integration may be preferable, which aligns with research aiming to overcome the unfeasible high volume of computations required by IIT [8,35].

We believe that the deepest problem in this case, which should be undertaken in further research, is to find an integrated notion that could account also for a temporal dimension. In particular, a formalism that could determine the dynamics of aggregation and disaggregation of elements into parts and viceversa. To be more specific; it may be the case that different elements, integrate into subsets at different timescales, therefore requiring a notion of temporal integration. To this end, though, first we would need to examine the specific problems of IIT regarding temporality [3,40], sketch potential solutions to them, and evaluate whether the theory really offers a solid framework in this regard.

## References

1. Abramova, K., Villalobos, M.: The apparent (ur-)intentionality of living beings and the game of content. Philosophia **43**, 651–668 (2015)
2. Aguilera, M., Di Paolo, E.: Integrated information in the thermodynamic limit. In: Proceedings of the ALIFE 2018: The 2018 Conference on Artificial Life. ALIFE 2018: The 2018 Conference on Artificial Life, Tokyo, Japan, pp. 113–120 (2018). https://doi.org/10.1162/isal_a_00030
3. Aguilera, M., Di Paolo, E.: Integrated information in the thermodynamic limit. Neural Netw. **114**, 136–149 (2019)

4. Albantakis, L., et al.: Integrated information theory (IIT) 4.0: formulating the properties of phenomenal existence in physical terms. PLoS Comput. Biol. **19**(10), e1011465 (2023). https://doi.org/10.1371/journal.pcbi.1011465
5. Balduzzi, D., Tononi, G.: Integrated information in discrete dynamical systems: motivation and theoretical framework. PLoS Comput. Biol. **4**(6), e1000091 (2008). https://doi.org/10.1371/journal.pcbi.1000091
6. Baltieri, M., Iizuka, H., Witkowsi, O., Sinapayen, L., Suzuki, K.: Hybrid life: integrating biological, artificial, and cognitive systems. WIREs Cogn. Sci. e1662 (2023). https://doi.org/10.1002/wcs.1662
7. Barandiaran, X.: Autonomy and enactivism: towards a theory of sensorimotor autonomous agency. Topoi **36**, 409–430 (2017). https://doi.org/10.1007/s11245-016-9365-4
8. Barret, A., Seth, A.: Practical measures of integrated information for time-series data. PLoS Comput. Biol. **7**(1), e1001052 (2011)
9. Beer, R.: Lost in words. Adapt. Behav. **28**(1), 19–21 (2020)
10. Beer, R., Di Paolo, E.: The theoretical foundations of enaction: Precariousness. Biosystems **223**(104823) (2023)
11. Beer, R.: Autopoiesis and cognition in the game of life. Artif. Life **10**, 309–326 (2004)
12. Beer, R.: The cognitive domain of glider in the game of life. Artif. Life **20**, 183–206 (2014)
13. Berlekamp, E., Conway, J., Guy, R.: Winning Ways for Your Mathematical Plays, vol. 2. Academic Press, New York (1982)
14. Cea, I.: Integrated information theory of consciousness is a functionalist emergentism. Synthese **199**, 2199–2224 (2021). https://doi.org/10.1007/s11229-020-02878-8
15. Clark, A.: Surfing Uncertainty: Prediction, Action and the Embodied Mind. Oxford University Press (2016)
16. Paolo, D., Burhmann, E., Barandarian, T., X.: Sensorimotor Life: An Enactive Proposal. Oxford University Press (2017)
17. Di Paolo, E.A., Thompson, E., Beer, R.D.: Laying down a forking path: tensions between enaction and the free energy principle. Philos. Mind Sci. **3**(2) (2022). https://doi.org/10.33735/phimisci.2022.9187
18. Froese, T.: To understand the origin of life we must first understand the role of normativity. Biosemiotics **14**, 657–663 (2021). https://doi.org/10.1007/s12304-021-09467-3
19. Froese, T., Taguchi, S.: The problem of meaning in ai and robotics: still with us after all these years. Philosophies **4**(14) (2019)
20. Gallagher, S.: Embodied and Enactive Approaches to Cognition. Cambridge University Press (2023)
21. Gallagher, S.: 4E cognition: historical roots, key concepts, and central issues. In: Newen, A., de Bruin, L., Gallagher, S. (ed.) The Handbook of 4E Cognition. Oxford University Press, Oxford (2018)
22. Gardner, M.: Mathematical games: the fantastic combinations of john conway's new solitaire game "life." Sci. Am. **223**, 120–123 (1970)
23. Hamon, G., Etcheverry, M., Chan, B., Moulin-Frier, C., Oudeyer, P.: Learning sensorimotor agency in cellular automata (2022). https://developmentalsystems.org/sensorimotor-lenia
24. Husbands, P.: Never mind the iguana, what about the tortoise? Models in adaptive behavior. Adapt. Behav. **17**(4), 320–324 (2009)

25. Hutto, D., Myin, E.: Evolving Enactivism. Basic Minds Meet Content. MIT Press (2017)
26. Hutto, D.D., Myin, E.: Deflating deflationism about mental representation. In: Smortchkova, J., Dołrega, K., Schlicht, T. (eds.) What Are Mental Representations? Oxford University Press, Oxford (2020)
27. Joshi, N., Tononi, G., Koch, C.: The minimal complexity of adapting agents increases with fitness. PLoS Comput. Biol. **9**(7), e1003111 (2013)
28. Kirchhoff, M., Froese, T.: Where there is life there is mind: In support of a strong life-mind continuity thesis. Entropy **19**(4), 169 (2017)
29. Lombardi, O., López, C.: What does 'information' mean in integrated information theory. Entropy **20**, 894 (2018)
30. Maturana, H., Varela, F.: Autopoiesis: the organization of the living. [De maquinas y seres vivos. Autopoiesis: la organizacion de lo vivo]. 7th edition from 1994. Editorial Universitaria (1973)
31. Maturana, H.: Everything said is said by an observer. In: Thompson, W.I. (ed.) Gaia: A Way of Knowing, pp. 65–82. Lindisfarne Press, New York (1987)
32. Maturana, H.: Biology of self-consciousness. In: Tratteur, G. (ed.) Consciousness: Distinction and Reflection, pp. 145–175. Bibliopolis, Naples (1995)
33. Mediano, P., Rosas, F., Bor, D., Seth, A., Barret, A.: The strength of weak integrated information theory. Trends Cogn. Sci. **26**(8), 646–655 (2022)
34. Mediano, P., Seth, A., Barret, A.: Measuring integrated information: comparison of candidate measures in theory and simulation. Entropy **21**(1), 17 (2018)
35. Mediano, P.A., Rosas, F., Carhart-Harris, R.L., Seth, A.K., Barrett, A.B.: Beyond integrated information: A taxonomy of information dynamics phenomena. arXiv:1909.02297 (2019). https://doi.org/10.48550/arXiv.1909.02297
36. Merker, B., Williford, K., Rudrauf, D.: The integrated information theory of consciousness: a case of mistaken identity. Behav. Brain Sci. **May**(19), 1–72 (2021)
37. Oizumi, M., Albantakis, L., Tononi, G.: From the phenomenology to the mechanisms of consciousness: integrated information theory 3.0. PLoS Comput. Biol. **10**(5) (2014). https://doi.org/10.1371/journal.pcbi.1003588
38. Pautz, A.: What is the integrated information theory of consciousness. A catalogue of questions. J. Conscious. Stud. **26**(1-2), 188–215 (2019)
39. Pocket, S.: Problems with theories that equate consciousness with information or information processing. Front. Syst. Neurosci. **8**(225) (2014). https://doi.org/10.3389/fnsys.2014.00225
40. Rodriguez, F., Husbands, P., Ghosh, A., White, B.: Frame by frame? A contrasting research framework for time experience. In: ALIFE 2023: Ghost in the Machine: Proceedings of the 2023 Artificial Life Conference, p. 75. MIT Press (2023). https://doi.org/10.1162/isal_a_00688
41. Rodriguez-Vergara, F., Husbands, P.: Is every cognitive phenomenon computable? Mathematics **14**(3), 535 (2026) https://doi.org/10.3390/math14030535
42. Rubner, Y., Tomasi, C., Guibas, J.: A metric for distributions with applications to image databases. In: Proceedings of the 1998 IEEE International Conference on Computer Vision, Bombay, India (1998)
43. Sayama, H.: Construction theory, self-replication, and the halting problem. Complexity **13**, 16–22 (2008). https://doi.org/10.1002/cplx.20218
44. Seth, A.K.: The cybernetic Bayesian brain: from interoceptive inference to sensorimotor contingencies. In: Metzinger, T., Windt, J.M. (eds.) Open MIND: 35(T). MIND Group, Frankfurt am Main (2015)
45. Stewart, J.: Cognition = life: implications for higher-level cognition. Behav. Proc. **35**, 311–326 (1996)

46. Tallis, R.: Freedom: An Impossible Reality. Agenda Publishing Limited (2021)
47. Thompson, E.: Mind In Life: Biology, Phenomenology and the Sciences of Mind. Harvard University Press, Cambridge (2007)
48. Thompson, E.: Sensorimotor subjectivity and the enactive approach to experience. Phenomenol. Cogn. Sci. **4**(4), 407–427 (2005)
49. Tononi, G.: An information integration theory of consciousness. BMC Neurosci. **5**(42) (2004)
50. Tononi, G., Koch, C.: Consciousness: here, there and everywhere? Philos. Trans. Roy. Soc. B **370**(1668) (2015). https://doi.org/10.1098/rstb.2014.0167
51. Tsuchiya, N., Taguchi, S., Saigo, H.: Using category theory to asses the relationship between consciousness and integrated information theory. Neurosci. Res. **107**, 1–7 (2016)
52. Varela, F.: Principles of Biological Autonomy. North Holland (1979)
53. Varela, F., Maturana, H., Uribe, R.: Autopoiesis: the organization of the living systems, its characterization and a model. Biosystems **5**(4), 187–196 (1974)
54. Varela, F., Thompson, E., Rosch, E.: The Embodied Mind: Cognitive Science and Human Experience. The MIT Press (1991)
55. Varela, F.: Structural coupling and the origin of meaning in a simple cellular automation. In: Sercarz, E.E., Celada, F., Mitchison, N.A., Tada, T. (eds.) The Semiotics of Cellular Communication in the Immune System. NATO ASI Series, vol. 23. Springer, Heidelberg (1988)
56. Varela, F.: Patterns of life: intertwining identity and cognition. Brain Cogn. **34**(1), 72–87 (1997)
57. Villalobos, M., Silverman, D.: Extended functionalism, radical enactivism and the autopoietic theory of cognition: prospects for a full revolution in cognitive science. Phenomenol. Cogn. Sci. **17**, 719–739 (2018)
58. Ward, M., Silverman, D., Villalobos, M.: Introduction: the varieties of enactivism. Topoi **36**, 365–375 (2017)
59. Weber, A., Varela, F.: Life after kant: natural purposes and the autopoietic foundations of biological individuality. Phenomenol. Cogn. Sci. **1**, 97–125 (2002)
60. Weng, L.: What is Wasserstein distance? (2017). https://lilianweng.github.io/posts/2017-08-20-gan/what-is-wasserstein-distance
61. Wiese, W., Friston, K.: Examining the continuity between life and mind: is there a continuity between autopoietic intentionality and representationality? Philosophies **6**(18) (2021). https://doi.org/10.3390/philosophies6010018
62. Wuensche, A.: The ghost in the machine: basins of attraction of random boolean networks. In: Artificial Life III: SFI Studies in the Sciences of Complexity, vol. XVI (1994)

# ALife in Complex Environments

# Characterization of Bow-Tie Zones in Unknown Systems

Carlo Uguzzoni[1] and Marco Villani[1,2(✉)]

[1] Department of Physics, Informatics and Mathematics, University of Modena and Reggio Emilia, Modena, Italy
marco.villani@unimore.it
[2] European Centre for Living Technology, Venice, Italy

**Abstract.** Many real-world systems, such as power grids and ecological networks, are vulnerable to failures that can trigger widespread disruptions. Understanding how the structure of a system influences its response to shocks is crucial to assessing its resilience: systems that can be represented by directed graphs allow a characterization of their resilience by identifying their Bow-Tie regions. When the system organization is unknown, mapping it becomes necessary, although challenging. Snowball sampling, a method that uses initial subject references to explore networks, is often used despite its limitations. Among the various alternatives, this work highlights how a depth-first approach to such sampling allows an early and effective characterization of the Bow-Tie regions - and therefore of the resilience characteristics - of the system under consideration.

**Keywords:** Complex networks · Bow-tie organization · resilience · snowball sampling · mapping approaches

## 1 Introduction

Many real-world systems are potentially subject to failures or external perturbations that cause disruptions and harmful consequences; for example, adverse events affecting the Internet, power grids, organizations, and natural systems [1–6]. The failure of a few nodes or links can cause a blackout in a power grid or the collapse of an ecological system [7, 8].

Therefore, understanding how the structure of a system determines its response to shocks is important, because it sheds light on its local and global resilience [9, 10] and how they change after a shock [11–14]. If the system can be represented by a directed graph (actually, many systems are), an effective way to understand whether the part affected by perturbations is a sensitive point of the system is to know its location within the bow-tie structure of that graph [15–17]. Node failures in the IN or SCC part can lead to major damage, while nodes belonging to other parts of the bow-tie organization represent less sensitive parts of the system [17, 18].

If the system organization is known, it is then possible to resort to such analysis. If, on the other hand, the organization is not known, or is only partially known, it is necessary to resort to mapping procedures. This can lead to numerous difficulties.


M. Bianchini et al. (Eds.): WIVACE 2025, CCIS 3001, pp. 147–157, 2027.
https://doi.org/10.1007/978-3-032-33185-4_11

In many cases in fact mapping actions are expensive (in terms of time, available resources, ...) or difficult to access through conventional means. A strategy frequently used in such cases is the so-called snowball sampling [19, 20]. Rather than selecting participants through random sampling or drawing from a complete list of individuals, researchers begin with a small number of initial subjects - often referred to as “seeds”, or “focal nodes” - who meet the criteria of the study. Starting from these initial nodes it is then possible to identify other nodes, belonging to their neighborhood. As new nodes are recruited, the sample expands according to a chain model, like a snowball that rolls and increases in size as it moves. The strength of snowball sampling lies in its ability to penetrate closed communities by exploiting the trust that exists between peers (someone is more likely to agree to participate if involved by a known and trusted contact), or to allow coordinated research action in unknown areas.

Despite its usefulness, snowball sampling has notable limitations. The most significant is that the sample is not random, and therefore the findings cannot be generalized to the entire population with statistical confidence.

In this work we are interested in observing the deformations induced in the size estimation of the bow-tie components of the system of interest (and therefore on the correctness of the estimate of its robustness) induced by different mapping strategies (including the snowball strategy), in case of limited resources (time, costs, ...). We will apply these strategies first to ensembles of known networks with different topologic architectures (Erdős–Rényi [21, 22] and scale-free systems [12, 23]), and then to the mapping of a real system (a network representing the global automotive system [17,18).

## 2 Bow-Tie Regions and Mapping Strategies

In their seminal 2000 work, Broder and colleagues proposed that the Web's directed graph organizes itself in a “bow-tie” shape, revealing a surprisingly structured macro-architecture. At the center lies the strongly connected core (SCC), a cluster of pages each reachable from every other via hyperlinks. Feeding into this core is the IN region: pages that link into the SCC but are not themselves reachable from it. Dually, on the other side lies the OUT region: pages reachable from the SCC but that have no path back into it. Beyond these, there exist tendrils and tubes—loosely connected nodes that lie outside the direct IN–SCC–OUT skeleton, either dangling off or linking across without traversing the core (Fig. 1). This bow-tie decomposition highlights how the Web is not a random jumble of pages but rather a highly stratified network, with a densely intertwined center and peripheral zones whose navigational roles differ dramatically [15, 16].

Any damage in the OUT zone, capable of transmitting to downstream nodes, which in turn can transmit the malfunction to further downstream nodes (etc. etc.), can cause an avalanche of perturbations that, however, is destined to remain blocked in the OUT zone, as this zone has no downstream contact with other zones. A similar damage initially within the SCC zone can completely invade it, even penetrating the OUT zone. Damage in the IN zone can propagate to the SCC zone, and from there expand throughout that zone until reaching the OUT zone. The Bow-Tie structure is therefore profoundly linked to the extent of the diffusion of perturbations (be they information or damages).

It is therefore important to be able to map these areas in the systems under consideration, whether they are known or yet to be explored (the situation studied in this work).

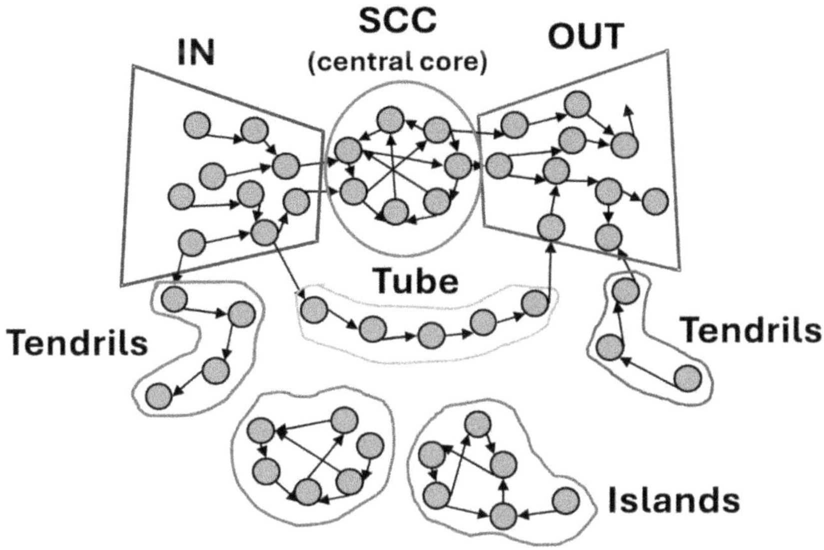


**Fig. 1.** The decomposition of a directed graph into its Bow-tie components. The IN, SCC, and OUT zones are highlighted, along with more bizarre zones such as Tubes (zones capable of directly connecting the IN zone to the OUT zone, without passing through the SCC zone) and Tendrils (zones that can exit from IN zone without reaching other zones, or that can reach the frontier of the OUT zone without allowing further information transmission).

There are two major groups of search/mapping strategies for unknown systems: breadth-first search (of which the Snowball strategy is a possible instance) and depth-first search [24].

Breadth-first search is an effective way to explore a largely unknown network: it consists of starting from a few known nodes (here called "focal nodes") and, in the case of directed networks, following the links originating from each node (first level). Once the search is complete, the iteration re-starts from the newly discovered nodes (subsequent levels). A schema of such a one-way search (called in the following BF1D) can be seen in Fig. 2, where also the (small for discussion reasons) network under examination is shown.

It is also possible to ignore the direction of the links and, having information about the incoming links at each node, to use both the outgoing and incoming links. A schema of this two-way breadth-first search (called in the following BF2D) involving the same previous network and the same previous focal points is shown in Fig. 3.

It can be noted that, despite the high number of discovered nodes, in the case of finite resources (and therefore of an early stop of the search) the two strategies could be unable to connect the different discovered sections of the network, thus leading to the incorrect Bow-Tie classification of the found nodes.

Depth-first search starts from a chosen source vertex, selects only one link, and iterates the procedure following a path as deep as possible. When there are no outgoing links in the current node, the method backtracks to the previous node that still has an unexplored link and starts again from there. The algorithm marks each visited node to

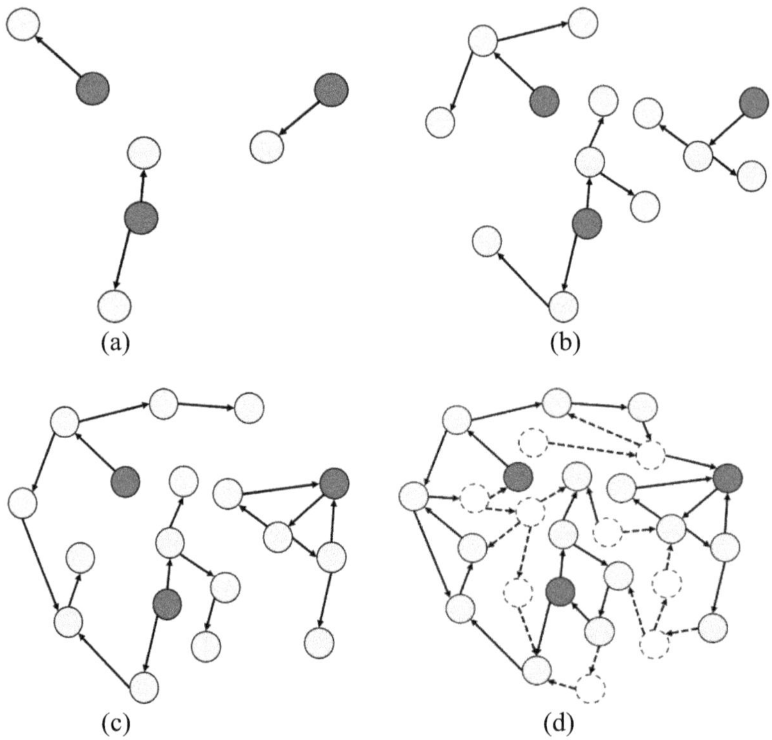


**Fig. 2.** One-directional breadth-first search. (a) The initial phase: the nodes immediately adjacent to the focal nodes, highlighted in red, are reached. (b) The second level and (c) the third level of search. (d) The complete network, with the nodes and links not yet traversed indicated by dashed lines.

avoid loops and repeated visits, ensuring that each reachable vertex is explored once. Depth-first search therefore tends to quickly explore long paths. If there are still unused focal nodes, the search can restart from one of them.

A schema of such a search following the link directions (called in the following DF1D) can be seen in Fig. 4, while Fig. 5 shows the same algorithm in the bidirectional case (where the link direction is ignored – DF2D).

We can note that:

- if the system under consideration is simply connected, the bidirectional algorithm always obtains a single connected component, which, in the case of no limits to the search, involves all focal nodes, without the need for restarts
- the depth of the discovered paths allows, in both cases (unidirectional and bidirectional search), a better sampling of the global characteristics of the system under consideration than breadth-first-based search strategies.

It is therefore possible that depth-first search may yield better results than breadth-first search, if we are interested in getting an idea of the size and location of bow-tie regions in a unknown systems. We will therefore apply these strategies first to sets of

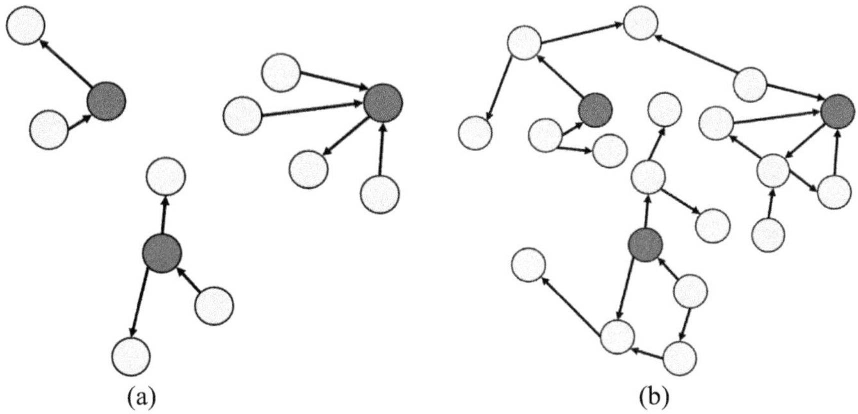


**Fig. 3.** Two-directional breadth-first search (snowball sampling). (a) The initial phase: the nodes immediately adjacent to the focal nodes, highlighted in red, are reached. (b) The second level of the search.

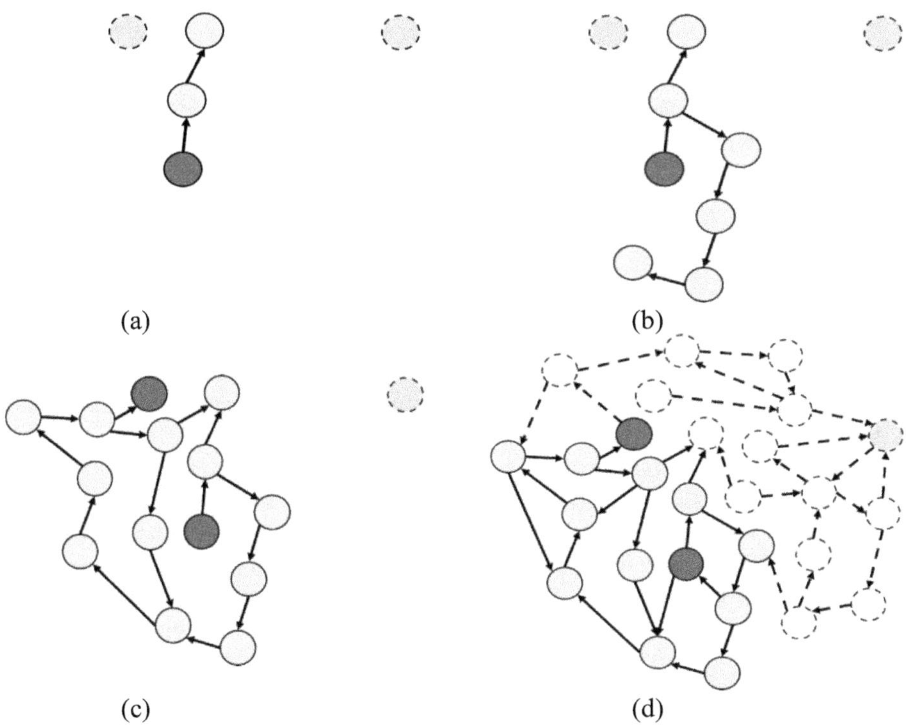


**Fig. 4.** One-directional depth-first search. (a) The initial phase: the first two steps of the method are shown. (b) The result is shown a few steps later, and (c) the situation where a number of nodes equal to the number of nodes found in Fig. 2c has been reached. (d) The complete network, with the nodes and links not yet traversed indicated by dashed lines.

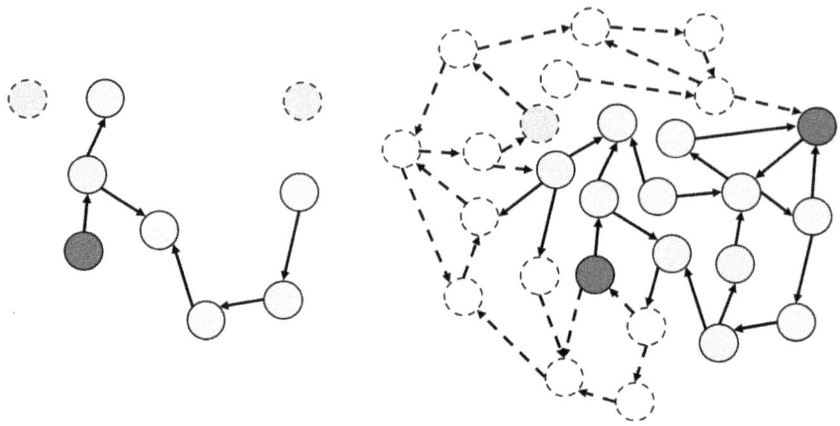


**Fig. 5.** Two-directional depth-first search. (a) The initial phase: the two steps of the method are shown. (b) The situation where a number of nodes equal to the number of nodes found in Fig. 2c has been reached. The nodes and links not yet traversed indicated by dashed lines.

artificial networks with different topological architectures (Erdős–Rényi and scale-free systems), and then to the mapping of a real system (a network representing the global automotive system [17, 18]).

## 3 Results

We then simulated the link discovery process aimed at identifying the typical size of bow-tie regions in ensembles of Erdős–Rényi (ER) and scale-free networks (BA) having $10^4$ nodes and average connectivity $k_m$ equal to 2, 3 up to 5; subsequently, we focused on networks with $k_m = 2$ (the situation with higher diversity). Each search started from a base of 10 focal nodes. In the following, we use ensembles of 100 networks: consequently, each point in the ensuing graphs represents the average of 100 values.

Figure 6 shows the number of nodes classified as belonging to each Bow-Tie region as the search progresses using the BF2D strategy – the approach also used during the automotive network study that we will show in the following, and which reproduces the typical snowball strategy. It is possible to notice that at each iteration the number of nodes discovered grows significantly (as expected), and that we get close to a correct quantification of the number of nodes belonging to the Bow-tie regions of greatest interest (IN, SCC and OUT) only very close to the end of the search. There is also an early and erroneous identification of peculiar regions (tubes, various types of tendrils), before correctly identifying their near absence. The research progress when applied to ER or BA systems is substantially similar, the IN and OUT regions being overestimated more in the BA case than in the ER case. The same happens with the BF1D strategy, with the added complication that during the search process the estimate of the various Bow-Tie regions changes significantly and abruptly (Fig. 7).

The behavior of the DF1D strategy appears much more regular (Fig. 8) and has the peculiarity that the IN region begins to be identified only after the identification of the

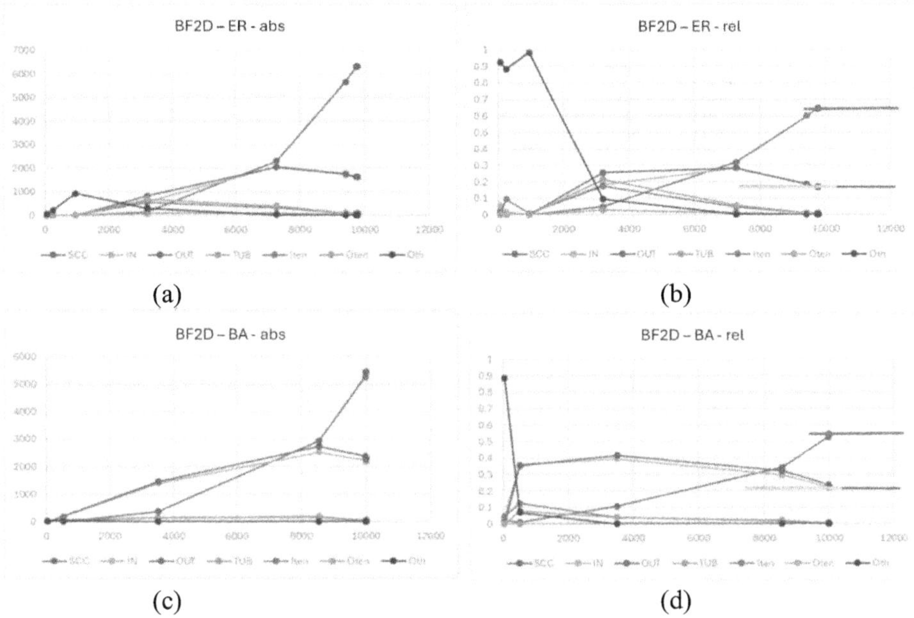


**Fig. 6.** BF2D strategy (snowball sampling). The first column shows the absolute number of nodes classified as belonging to each Bow-Tie region as the search progresses, the second column shows the relative fraction (the denominator being the total number of nodes identified). The figures in the second column also show the correct values for the IN (orange), SCC (blue) and OUT (green) regions. The first row shows the case in which the system under consideration has an ER architecture, the second row the case in which the architecture is BA. The graphs present a midpoint at each level, until the final data is shown once all reachable nodes have been identified.

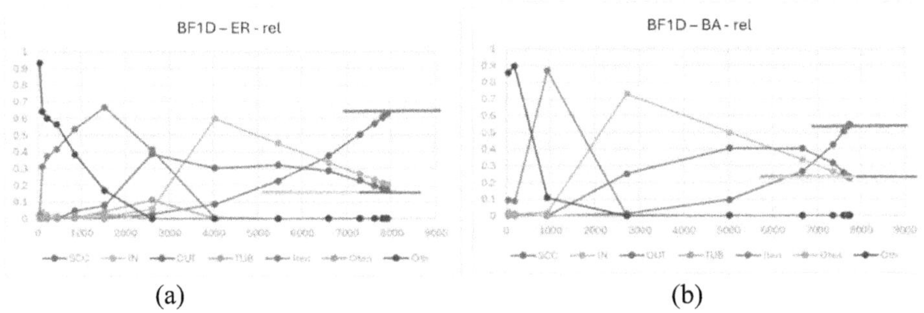


**Fig. 7.** BF1D strategy. Evaluation of the relative size of Bow-Tie zones in systems with (a) ER architecture and (b) BA architecture.

SCC and OUT regions has been completed. In fact, in the case of a one-directional strategies, the identification of nodes belonging to the IN region is possible only by starting from other IN nodes: in our case, it is necessary to resort to "restarts" from focal nodes not reached by the current search.

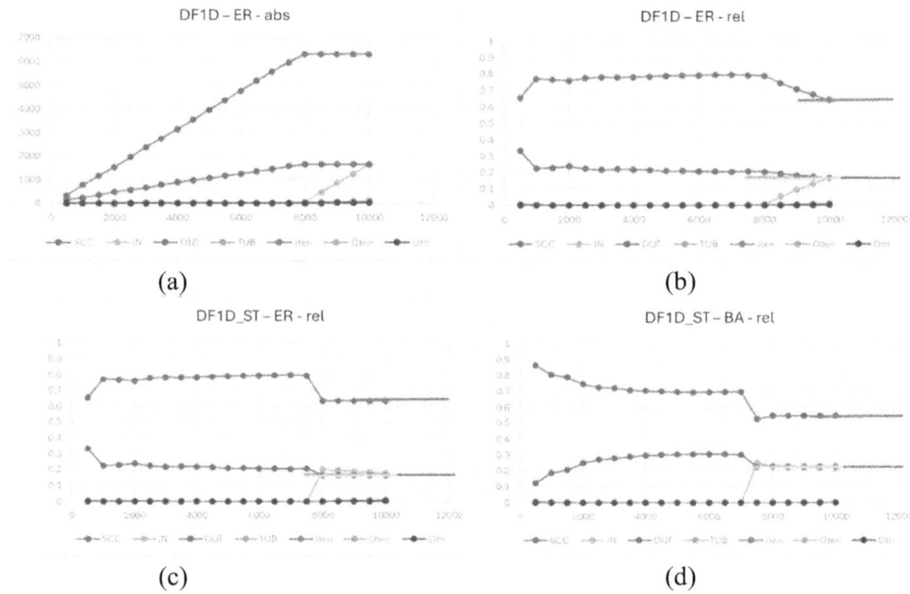


**Fig. 8.** DF1D Strategy. (a) Absolute size of each Bow-Tie region as the search proceeds for systems with ER architecture. (b) Relative size of each Bow-Tie region as the search proceeds for systems with ER architecture. (c) As in (b) but estimating the size of the IN region as the difference between the total number of nodes and the sum of the SCC and OUT regions (the difference being divided by the total number of nodes identified – strategy DF1D_ST). The application of this strategy allows an early estimation of the IN region (d) As in (c), for systems with BA architecture. The graphs present a midpoint every 500 steps.

Conversely, since in the case of ER and BA networks particular Bow-Tie zones (tubes, various types of tendrils) are practically nonexistent, assuming that—once the identification of the SCC and OUT nodes has been completed—all the remaining nodes belong to the IN region is a good approximation. This stratagem (called in the following DF1D_ST) is obviously only applicable when the total number of nodes present in the system is known. The research progress when applied to ER or BA systems is substantially similar, the OUT region being overestimated more in the BA case. The stratagem BF1D_ST is applicable also in BA systems (Fig. 8d).

In order to obtain an evaluation of the size of the IN region before completing the identification of the SCC and OUT regions, we need to use the DF2D strategy, whose progress is shown in Fig. 9.

In summary, the best breadth-first strategy is BF2D, while the best depth-first strategy is DF1D for identifying the SCC and OUT regions only, and DF2D for an early identification of the IN region. When DF1D_ST can be used, this approach may be better than the DF2D strategy.

Comparing the best strategies of the two search types, BF2D and DF1D (or DF1D_ST), we can see that BF2D requires knowledge of almost all the nodes in the system under consideration to correctly quantify the Bow-Tie regions (both in the ER

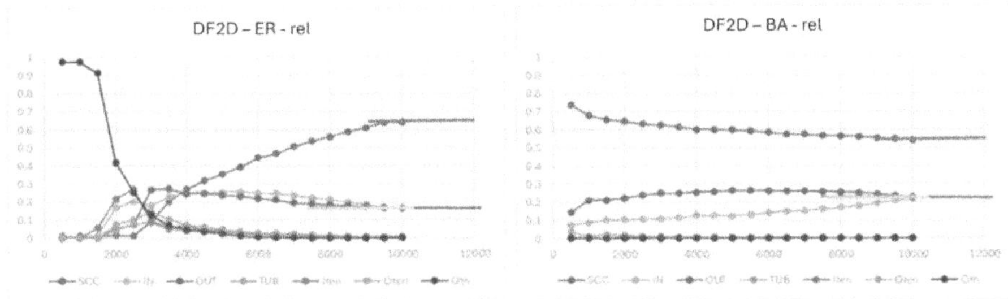


(a) (b)

**Fig. 9.** DF2D strategy. Evaluation of the relative size of Bow-Tie zones in systems with (a) ER architecture and (b) BA architecture.

and BA cases – Fig. 6b and 6d), while DF1D (DF1D_ST) requires less than 80% (Fig. 8c and 8d).

Moreover, we can note that the DF1D strategy provides a stable estimate of the SCC and OUT regions (slightly overestimated, but in a systematic—and therefore "known"—way) starting from knowledge of only 25%–30% of the nodes in the case of BA systems, and 10%–15% of the nodes in the case of ER systems (Figs. 8d and 8c). A similar observation can also be made for the BF2D strategy (as far as regards IN and OUT regions), but with a continuously varying estimate of the SCC region (Figs. 6b and 6d), and the presence of significant false positives as far as regards the Tube and Tendril regions.

Indeed, the identification of bow-tie regions by breadth-first search suffers from several difficulties, differently depending on whether the architecture of the system under consideration is ER or scale-free. Starting with few focal nodes (few compared to the number of nodes to be discovered) implies that the set of nodes discovered in the first search levels is not even connected: the verification of the simple connection occurs earlier for scale-free systems, thanks to the connective action of the hub nodes (Fig. 6b and 6c). There is an overestimation of the relative size of the Tendril regions and a strong underestimation of the relative size of the SCC region, a situation that persists almost until the complete classification of all the nodes of the system. The estimate of the relative size of the IN region is early (better in the case of scale-free systems), while the identification of the OUT region is comparable to the case of depth-first search. Increasing the number of initial focal nodes, bringing them to 50 and then to 100, does not significantly change these observations (data not shown).

Finally, breadth-first strategies are not flexible: in order to obtain consistent results, one would need to proceed from level to level, and at each level the number of nodes to be processed increases significantly. Conversely, it is possible to stop a depth-first strategy (or obtain estimates) at each step.

It is possible to apply the strategies thus defined to real systems. [17] describes network based on supplier-customer relationships that describes the global automotive economic system, composed of 3323 companies covering 135 industrial sectors and 16 geographical areas, connected by 11182 mutual commercial relationships [17, 18]. The identified network was constructed using the snowball approach, starting from 164 focal nodes; following the customer-supplier and supplier-customer relationships for two levels of depth (the suppliers and customers of the customers of each focal node were identified, and similarly the suppliers and customers of the suppliers of each focal node). Regarding connectivity, the network exhibits distributions that follow power laws [12], but the growth process could hardly be described as a pure Barabasi-Albert [23].

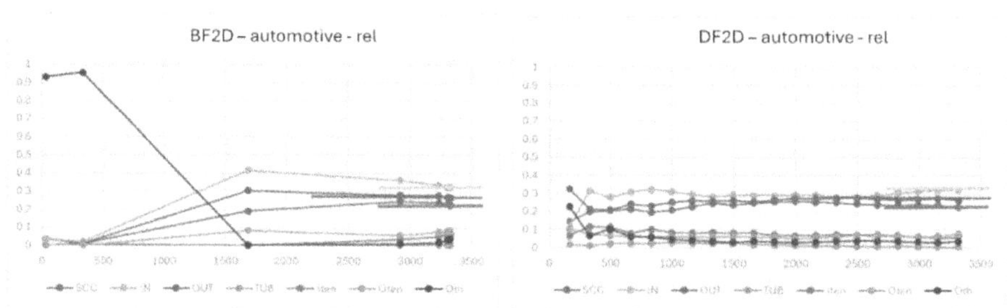


(a) (b)

**Fig. 10.** Identification of Bow-Tie zones in the automotive network. (a) BF2D strategy. (b) DF2D strategy.

In Fig. 10 it can be observed that using a depth-based strategy the identification of the Bow-Tie regions occurs significantly earlier than using a breadth-based strategy (in this case we applied the "weaker" DF2D strategy, because we were interested in an early identification of the IN region as well). It is also possible to note that the DF2D strategy allows an early identification of the Tube and Tendril regions.

## 4 Conclusions

If one is interested in characterizing the robustness of a system subject to perturbations, a common procedure is that of representing it by means of a graph: if the graph is directed, the representation can be used to identify its Bow-Tie areas, which are related to the robustness characteristics of the system. However, if the internal relationships of the system are not known and must be identified (a very frequent case), the mapping method can significantly affect the effectiveness of the procedure.

In this work, we show that a depth-first approach allows a better efficiency in identifying Bow-Tie areas than a breadth-first approach, allowing an early characterization of the system (in particular with regards to the SCC and OUT areas). We have shown this property in artificial systems characterized by Erdős–Rényi and scale-free architectures (built using the algorithm proposed by Barabasi and Albert). We also described some regularities of the two search strategies and their consequences for identifying Bow-Tie regions.

Finally, we applied the best methods to a real-world case (a network representing 3323 nodes and 11800 relationships in the global customer-supplier system in the automotive industry), finding that the depth-first approach allows for a better and significantly earlier characterization of Bow-Tie regions than the breadth-first approach.

**Acknowledgments.** This research was partially funded by Università degli Studi di Modena e Reggio Emilia (FAR2025 project of the Department of Physics, Informatics and Mathematics).

## References

1. Albert, R., Jeong, H., Barabási, A.L.: Error and attack tolerance of complex networks. Nature **406**, 378–382 (2000)
2. Pastor-Satorras, R., Vespignani, A.: Epidemic spreading in scale-free networks. Phys. Rev. Lett. **86**, 3200–3203 (2001)
3. Gao, J., Barzel, B., Barabasi, A.L.: Universal resilience patterns in complex networks. Nature **530**, 307–312 (2016)
4. Shai, P., Porter, M.A., Pascual, M., Kefi, S.: The multilayer nature of ecological networks. Nat. Ecol. Evol. **1**(4), 0101 (2017)
5. Yang, Y., Nishikawa, T., Motter, A.E.: Small vulnerable sets determine large network cascades in power grids. Science **358**, N6365 (2017)
6. Zhao, J.H., Zhou, H.J., Liu, Y.Y.: Inducing effect on the percolation transition in complex networks. Nat. Commun. **4**, 2412 (2013)
7. Dunne, J.A., Williams, R.J.: Cascading extinctions and community collapse in model food webs. Philos. Trans. Roy. Soc. B: Biol. Sci. **364**(1524), 1711–1723 (2009)
8. Buldyrev, S.V., Parshani, R., Paul, G., Stanley, H.E., Havlin, S.: Catastrophic cascade of failures in interdependent networks. Nature **464**, 1025–1028 (2010)
9. Serra, R., Villani, M., Graudenzi, A., Kauffman, S.A.: Why a simple model of genetic regulatory networks describes the distribution of avalanches in gene expression data. J. Theor. Biol. **246**(3), 449–460 (2007)
10. Villani, M., La Rocca, L., Kauffman, S.A., Serra, R.: Dynamical criticality in gene regulatory networks. Complexity **2018**, 1–14 (2018)
11. Cohen, R., Erez, K., Ben-Avraham, D., Havlin, S.: Resilience of the internet to random breakdowns. Phys. Rev. Lett. **85**(21), 4626–4628 (2000)
12. Strogatz, S.H.: Exploring complex networks. Nature **410**(6825), 268–276 (2001)
13. Watts, D.J.: A simple model of global cascades on random networks. Proc. Natl. Acad. Sci. **99**(9), 5766–5771 (2002)
14. Tieri, P., et al.: Network, degeneracy and bow tie. Integrating paradigms and architectures to grasp the complexity of the immune system. Theor. Biol. Med. Modell. **7**(1), 1–16 (2010)
15. Broder, A., et al.: Graph structure in the web. Comput. Netw. **33**(1–6), 309–320 (2000)
16. Fujita, Y., Kichikawa, Y., Fujiwara, Y., Souma, W., Iyetomi, H.: Local bow-tie structure of the web. Appl. Netw. Sci. **4**, 15 (2019)
17. Flori, E., Zhu, Y., Paterlini, S., Pattarin, F., Villani, M.: Spread of perturbations in supply chain networks: the effect of the bow-tie organization on the resilience of the global automotive system. In: De Stefano, C., Fontanella, F., Vanneschi, L. (eds.) Artificial Life and Evolutionary Computation, WIVACE 2022. CCIS, vol. 1780. Springer, Cham (2023)
18. Flori, E., Paterlini, S., Pattarin, F., Villani, M.: Evaluation of the pandemic impact on global automotive supply chain through network analysis. J. Econ. Interact. Coord. (2025)
19. Goodman, L.A.: Snowball sampling. Ann. Math. Stat. **32**(1), 148–170 (1961)
20. Brace-Govan, J.V.F.: Issues in snowball sampling: the lawyer, the model and ethics. Qual. Res. J. **4**(1), 52–60 (2004)
21. Erdős, P., Rényi, A.: On random graphs. I (PDF). Publicationes Mathematicae **6**(3–4), 290–297 (1959)
22. Bollobás, B.: Random Graphs, 2nd edn. Cambridge University Press (2001)
23. Albert, R., Jeong, H., Barabási, A.-L.: Diameter of the world-wide web. Nature **401**(6749), 130–131 (1999)
24. Leiserson, C.E., Cormen, T.H.: Introduction to Algorithms. MIT Press (1991)

# Toward a Gaian Synthetic Approach Techno-Apoptosis as a Catalyst for Sustainable Robotics

Luisa Damiano(✉) and Antonio Fleres

IULM University, 20143 Milan, Italy
{luisa.damiano,antonio.fleres}@iulm.it

**Abstract.** In response to the ongoing ecological crisis, this paper offers a methodological and theoretical proposal for reimagining scientific and technological innovation by placing sustainability at its core—not as an external constraint, but as a generative principle for technoscientific development. We focus in particular on the potential of the synthetic approach, grounded in the epistemological principle of *understanding by building*, as a foundation for modelling and designing sustainable robotic systems. By tracing the development of this approach, we propose a novel trajectory for its evolution: a *Gaian turn* in the synthetic approach, aimed at creating robotic technologies that, by modelling natural self-regulating dynamics, can integrate into—and actively contribute to—the self-regulation of their ecological contexts. As an exploratory contribution to this *Gaian synthetic approach*, we introduce the framework of *techno-apoptosis*. Building on exploratory foundational work [11], we develop this framework as a tool to deepen the scientific understanding of apoptosis in order to inform the design of robotic systems capable of self-dismantling, modular reuse, and integration into circular techno-ecologies. This programmatic proposal outlines not only an epistemological, methodological, and theoretical architecture, but also a principled approach to reimagine synthetic modelling as a practice capable of engaging with finitude, transformation, and systemic viability in different domains – from artificial life to synthetic biology.

**Keywords:** Gaian Synthetic Approach · Techno-Apoptosis · Sustainable (Chemical and Electromechanical) Robotics

## 1 Introduction

Human impact on the Earth system has reached unprecedented levels of intensity and scale. The structural unsustainability of the dominant techno-industrial paradigm calls for a fundamental rethinking of technological innovation. Among today's most transformative and fast-evolving domains, robotics plays a critical role. Indeed, it can either deepen systemic fragilities or support transitions toward sustainability—if, and only if, its design principles are radically reconfigured.

Most current approaches to sustainable robotics prioritize energy efficiency, material optimization, or passive end-of-life management strategies. While valuable, these remain


M. Bianchini et al. (Eds.): WIVACE 2025, CCIS 3001, pp. 158–170, 2027.
https://doi.org/10.1007/978-3-032-33185-4_12

largely technical responses that do not question the epistemological foundations of system design. This paper proposes a deeper reorientation, which embeds sustainability into the generative logic of modeling and construction.

We contribute to this reconfiguration by proposing an evolutionary shift in the synthetic approach: a *Gaian turn*, inspired by the Gaia hypothesis formulated by Lovelock and Margulis [18, 20], which envisions the Earth as a complex, self-regulating system (see §4.3). This reorientation grounds synthetic modeling in principles of ecological embeddedness, systemic self-organization, and planetary viability. Within this Gaian paradigm, we introduce the framework of "techno-apoptosis" [11]: the programmed self-dismantling of robotic systems as a concrete strategy for sustainability and regenerative design.

What follows is an exploration of how the synthetic approach, understood not only as an epistemological method but also as a systemic and enactive mode of inquiry [9], can offer a robust foundation for rethinking robotic systems as agents embedded within techno-ecological landscapes. Within this perspective, the techno-apoptotic framework we develop responds to the ecological crisis not through linear solutions, but by embracing a circular and anticipatory logic of integration, decay, and renewal—reframing dismantling not as failure, but as a systemic function.

## 2 General Scope and Methodological Framing

This paper presents a methodological and theoretical contribution, grounded in the history and epistemology of the synthetic approach, and oriented toward its possible evolution in response to sustainability challenges. It does not introduce new experimental data or technical prototypes. Instead, it adopts an epistemology grounded in complex systems thinking and synthetic modeling to reformulate how we conceive the lifecycle, function, and role of artificial systems within broader ecological and organizational contexts.

The methodology adopted in this paper is speculative in a precise, epistemologically grounded sense. We draw on the history of ideas, philosophical inquiry, and analogical reasoning to articulate a space of possibility for future experimental, scientific, and technological developments. The aim is not to propose definitive solutions, but to frame relevant questions and address them by outlining potential directions for synthetic inquiry.

In line with this positioning, the following research questions are identified as central to the inquiry.

*Epistemological.* How can the synthetic approach, grounded in *understanding by building*, be reoriented through a Gaian turn to incorporate principles of bounded systemic conditions, ecological interdependence, and regenerative transformation?

*Methodological.* What does it mean to articulate a Gaian reconfiguration of the synthetic approach, and how does this affect its principles, scope, and implementation?

*Theoretical.* How can techno-apoptosis—a biologically grounded principle of programmed self-dismantling—serve as a theoretical and modeling framework for implementing the Gaian synthetic approach in the design of sustainable robotic systems?

*Transdisciplinary.* How can this Gaian methodological reorientation foster new synergies among robotics, synthetic biology, artificial life, and complex systems sciences in addressing planetary viability?

Structured around these questions, our contribution proposes a shift in scientific modeling and technological design, grounded in the epistemology of the synthetic approach and engaged with the ecological dimension of techno-scientific developments. In doing so, it promotes a methodological framework that links the synthetic exploration of life and cognition to principles of transformation, interdependence, and ecological viability, understood as dimensions of sustainability.

# 3 The Evolution of Synthetic Modelling

## 3.1 Building a Synthetic Epistemology on "Machines that Think"

Born in the early 20th century at the intersection of engineering, biology, and experimental psychology, the "synthetic method" marked a turning point in the evolution of science, particularly regarding the scientific study of cognition and life. Its origins can be traced to what Cordeschi [4] called "proto-cybernetics", an exploratory tradition that laid the foundations for the cybernetic movement. In particular, the work of John Hammond Jr., Benjamin Miessner, and Jacques Loeb is pivotal. These early pioneers built machines—such as the "electric dog", inspired by Loeb's theory of tropisms—that were conceived not as *metaphors*, but as *experimental models* designed to operationalize theoretical hypotheses in behavioral biology. The core idea was clear: to build in order to understand—or, as revived by embodied AI's pioneers in the 1990s, "understanding by building" [22].

One of the earliest and most detailed articulations of the synthetic method appears in the work of engineer Thomas Ross, who designed and experimentally explored electromechanical "learning machines". In a 1933 article describing these devices as "machines that think", Ross introduced a maze-solving setup inspired by behaviorist learning theories, featuring memory-based mechanisms. He named this experimental strategy the "synthetic method," framing it as a way to advance scientific understanding of natural cognition through the construction and empirical study of electromechanical devices [24].

Ross's contribution anticipated core elements of this emerging scientific method: (1) constructing material models based on hypotheses about how target processes are generated in nature; (2) using these models as testbeds to replicate and investigate those processes; (3) evaluating the operational power of the hypotheses through comparison between artificial and natural processes. Ross explicitly directed this approach not at structural imitation, but at functional reproduction, thereby foreshadowing two key, diverging trajectories in synthetic modeling: the functionalist logic of multiple realizability, and the emergent logic of complementarity between structure and functional organization. The former interprets synthetic systems as instantiations of functions that can be realized across diverse physical substrates. The latter emphasizes that cognitive and biological processes cannot be separated from the material and organizational conditions through which they emerge [9].

Building on the synthetic practice of Ross and other early pioneers, cybernetics and its scientific developments progressively extended the method—first conceptually, then operationally—to include computational and chemical implementations. This evolution gave rise to what is now termed the "synthetic approach": a cross-media, cross-disciplinary approach rooted in the principle of *understanding by building*, which supports exploratory research on life and cognition across computational and embodied AI, artificial life, robotics, and synthetic biology.

The progressive emergence of the synthetic approach marked a key epistemological shift in science, fostering the development of what we call a *synthetic epistemology*.

## 3.2 The Synthetic Approach as an Epistemological Innovation

In the foundational phase of cybernetics, figures such as Kenneth Craik, Warren McCulloch and Walter Pitts, and Arturo Rosenblueth, Norbert Wiener, and Julian Bigelow (1943) rigorously articulated the synthetic approach as a necessary complement to the classical analytical approach of scientific rationality. While the analytical method dissects and decomposes to reveal elementary structures, the synthetic method builds operational systems to investigate complex emergent dynamics, based on the experimental testing of hypotheses about underlying generative mechanisms. These contributions not only legitimized the role of synthetic modeling in scientific inquiry, but also advanced its implementation by articulating a more structured methodological cycle—an evolution of the experimental strategy already characterizing Ross's early work.

In continuity with protocybernetic synthetic practices, this cybernetic formulation of the synthetic approach specified a three-phase operational cycle. *(1.) Construction.* Designing an artificial system that can be used to synthetically generate the processes under investigation, based on theoretical assumptions about their generative mechanisms. *(2.) Environmental Embedment.* Integrating the synthetic model within a natural, artificial, or mixed environment consistent with the theoretical framework, enabling system–environment interactions that give rise to the emergence of the target processes. *(3).Comparative Evaluation.* Assessing the behavior of the artificial system in relation to that of its natural counterpart, to evaluate the initial hypotheses and the model's operational explanatory power [8].

This cyclic articulation—still guiding contemporary synthetic practices that involve robotic, computational, or chemical systems—emphasizes the central role of system–environment interaction and experimental comparison as integral to theory development and refinement. Its epistemological ambition lies in the productive use of divergences between natural and artificial dynamics—not as failures, but as *diagnostic mismatches.* These discrepancies, when not caused by material flaws in the model, are treated as meaningful differences that expose limitations in the original hypotheses and support their refinement or broader theoretical reformulation. Methodologically, this enables new iterations of the synthetic cycle, with a twofold aim: to improve scientific understanding of biological and/or cognitive processes, and to generate advanced technologies informed by that enhanced understanding.

It is through this iterative process that the synthetic method gives rise to a new synthetic epistemology and a new way of doing science, diverging from the tradition of classical science. Here, the goal is not to predict a process, but to reconstruct its unpredictable complexity. The strategy at play is not control, but generation. Indeed: the object of study is not pre-given, but constructed; the observer is not neutral, but the creator of the system and its behaviors; the model is not a transparent, objective representation, but a generator of phenomena that would not exist without it; explanation is no longer framed as deduction from general laws, but as an operational account developed through the production of dynamics under specific constraints and the observation of what those conditions bring forth [8].

On these grounds, the synthetic approach marks a transition in scientific practice from representationalism to constructivism. That is: from a detached, external science to an interpretive and creative mode of inquiry, in which classical objective descriptions are replaced by artifacts built by the observer as "revealing machines"—material constructs that, by surprising their creator with unpredictable behaviors, can reorganize the scientific understanding of life and cognition [23].

This shift is deeply rooted in the epistemology of complex systems sciences. In line with it, the synthetic approach operationalizes key principles of complex systems and related scientific inquiry—such as emergence, self-regulation, circular causality, trans-disciplinary coupling, and open-ended research—by integrating them into a scientific practice: a way of doing science that, rather than assuming total intelligibility, engages with systems that exceed the observer's control—complex systems that transform the very limits of knowledge into thresholds for innovation.

# 4 Toward a Gaian Paradigm for the Synthetic Approach

## 4.1 Exploring the Possible

Given this epistemological orientation, the synthetic approach finds one of its most distinctive traits in a *heuristic of the possible*, suited to engaging with systems that resist full intelligibility. From the early work of Thomas Ross and Nicolas Rashevsky [4] to contemporary applications [3], this heuristic foregrounds the value of constructing multiple, non-identical artificial realizations of natural phenomena—models that, while differing in structure or medium, are designed to enact similar generative logics under specific material constraints. The aim is not replication, but the exploration of how different configurations can give rise to analogous dynamics.

This heuristic not only expands the space of possible artificial systems, but also recognises the material and ecological limits that enable their construction. Each synthetic model unfolds within material, temporal, and ecological limits—its finitude as a condition of generativity. Recognizing these boundaries transforms possibility itself into a generative constraint: a way to design systems that anticipate their own transformation and reintegration. In this sense, the heuristic of the possible becomes a bridge between open-ended exploration and the deliberate incorporation of limits—the conceptual ground on which the techno-apoptotic framework will later stand (Sect. 5).

In synthetic epistemology, this approach resonates deeply with autopoietic and enactive perspectives, in which cognition arises through embodied action and organism–environment coupling. Building on this framework, recent taxonomies of synthetic models [9] have shown how they function not merely as experimental tools but also as enactive spaces of knowledge production, where explanation is constructed through interaction.

The partial and situated nature of our knowledge thus becomes a source of generative insight—a way to design synthetic systems that participate in ecological dynamics rather than merely replicate them.

This continuity between heuristic possibility and bounded design provides the conceptual bridge toward techno-apoptosis, developed in Sect. 5.

### 4.2 Robotics as a Synthetic Frontier

Among the domains where the synthetic approach has most deeply evolved, robotics stands out as a particularly fertile terrain. Since the 1990s, the embodied cognition paradigm—pioneered by Rodney Brooks, Rolf Pfeifer, and Luc Steels [22, 28]—has focused on embodied agents capable of autonomous interaction with the physical world. Robotics, in this context, does not merely simulate behaviour but enacts it: robots become epistemic artefacts whose dynamics emerge through agent–environment coupling. In this way, robotics exemplifies—and extends—the synthetic principle of *understanding by building*, engaging not only with material constraints but increasingly with ecological ones.

This evolution has generated a spectrum of research lines—evolutionary, swarm, soft, and chemical or *wetware* robotics—that collectively expand the reach of synthetic modelling across materials and scales. Evolutionary and swarm robotics explore how self-organization and distributed intelligence arise from local interactions [1, 10], while soft and morphological robotics investigate how physical structure and material properties shape behaviour [15]. Beyond electromechanical embodiments, emerging directions in *chemical or wetware robotics* [6, 26, 27] extend this logic to prebiotic and molecular domains, where cognition and regulation are explored through reticular chemical self-organization. These developments further expand the horizon of artificial embodiment, aligning the synthetic exploration of life and cognition with planetary ecological processes.

Across these heterogeneous expressions, robotics appears less as a single technology than as a synthetic mode of inquiry into embodied and ecologically embedded cognition. This continuity across electromechanical, soft, and chemical substrates outlines the conditions under which *techno-apoptosis* can be conceived not as a biological metaphor but as an operational principle—one that redefines dismantling and reuse as modes of systemic participation rather than loss.

Contemporary robotics thus constitutes both an epistemic laboratory and a conceptual frontier for rethinking the synthetic approach in light of sustainability, circularity, and ecological integration. This opens the way for the Gaian reconfiguration developed in the following section.

### 4.3 A Gaian Paradigm for Synthetic Research

As anticipated, the synthetic approach typically pursues two interconnected yet distinct aims: the epistemological ambition of *understanding by building*, and the engineering drive to construct artificial systems inspired by natural dynamics and mechanisms. Today, as sustainability emerges as a condition for systemic viability, these two dimensions reveal a new potential convergence. Through its performative, iterative, and materially embedded logic, the synthetic approach appears increasingly suited—and ready—for a fundamental epistemological evolution, one characterized by an explicit ethical orientation.

We propose this evolution as a shift toward a *Gaian paradigm* of the synthetic approach: a reconfiguration of synthetic scientific exploration grounded in the principles of ecological embeddedness, systemic self-organization, and planetary viability. Inspired by the Gaia hypothesis and its extensions in Lovelock's *Novacene* [19], we use the term *Gaian* to denote a systemic vision of the Earth as a complex whole in which biological, geological, and technological processes co-regulate. A whole in which an "electronic biosphere" is populated by autonomous machines that, far from being threats, support planetary self-regulation as participants in systemic resilience. In this view, artificial systems are not merely tools of control but potential co-regulators of ecological balance; and modelling them becomes a gesture of alliance with the biosphere—an effort to generate not only operational knowledge but relational viability.

Unlike current evolutionary or swarm-robotics programs that focus on adaptive optimization and the autonomous evolution of robotic units, the Gaian synthetic approach introduces a different kind of innovation. Its novelty lies not in designing new control mechanisms, but in reconfiguring the *epistemic orientation* of design itself: from *evolution-as-optimization* to *auto-eco-regulation-as-participation*—that is, a mode of design in which systems regulate themselves through their ecological interactions.

We express this Gaian reconfiguration through a reformulation of the traditional triadic epistemological cycle. (1) *Constructive Modelling*. Systems are designed not merely to reproduce target natural dynamics, but to express principles such as modularity, regeneration, and designed decay. Their architecture incorporates not only generative hypotheses about the processes under investigation, but also ecological hypotheses, such as the capacity for self-dismantling and sustainable reintegration as forms of embedded self-regulation. (2) *Ecological Embedment*. Synthetic models are situated within real or artificial environments, where their activity intersects with systemic dynamics. The aim is not just performance, but adaptive integration within local ecologies, material flows, and energy balances. (3) *Systemic Evaluation and Renewal*. Evaluation shifts from functional efficiency to systemic contribution: artificial systems are assessed for their capacity to support, reflect, or regenerate the viability of the broader systems in which they operate. The outcome is redirection—a new iteration of design and construction.

This triadic cycle grounds a new class of experimental frameworks that integrate synthetic systems into ecosystemic viability. Among them, we introduce *techno-apoptosis* as a particularly radical expression: a synthetic strategy oriented toward self-dismantling, component reuse, and circular regeneration.

In this sense, the *Gaian turn* in the synthetic approach does not propose new mechanisms of control, but new modes of participation—an epistemological and ethical reconfiguration of how artificial systems coexist with the biosphere. It redefines sustainability as a methodological commitment and a moral horizon: to understand by building systems that can, in turn, sustain.

# 5 From Apoptosis to Techno-Apoptosis Introduction

## 5.1 Philosophical Foundations and General Framing of Techno-Apoptosis

The epistemological and ethical reconfiguration proposed by the Gaian paradigm of the synthetic approach finds philosophical grounding in Peter Sloterdijk's "homeotechnological turn" [17, 29]. In contrast to *allotechnics*—a technological paradigm based on extraction, control, and separation from nature—*homotechnics* promotes an integrative vision of technology based on non-dominant and co-operational logics [25], anticipating a planetary homeotechnics consistent with the Gaian paradigm developed here. It invites the design of artificial systems aligned with principles observed in natural systems: cyclicality, self-regulation, co-evolution, and regenerative adaptation [2]. Technologies, in this view, are not instruments of domination but agents of ecological participation.

This perspective transforms the role of synthetic modeling: artificial agents are not conceived as discrete tools but as nodes embedded in dynamic ecologies. A homotechnical approach to the synthetic approach thus enables a shift from mere functionality to ethical contribution: designing systems that not only adapt to but actively support and regenerate their environments. As Sloterdijk suggests, the future of technology lies not in mastery over nature, but in consciously inhabiting its logic.

A further grounding for our proposal emerges in Edgar Morin's complex systems theory, particularly in his interpretation of apoptosis [21]. Morin reframes programmed cell death not as failure, but as "functional sacrifice": a form of *systemic altruism* in which individual elements self-eliminate to sustain collective coherence, viability and adaptability. This principle is observable not only in biology but also in social systems, where forms of "social apoptosis", as seen in eusocial colonies like *Apis mellifera*, support collective resilience and adaptability through mechanisms of organization and self-regulation [14]. It is on this background that, based on previous groundwork [11], we introduce *techno-apoptosis* as a concrete and radical articulation of Gaian synthetic modeling. This approach takes inspiration from biological apoptosis not as a metaphor, but as a design principle, construed as the capacity of artificial systems to self-dismantle, repurpose their components, and reintegrate into systemic material cycles. These features are not failures to be avoided, but core design goals. They express finitude, modularity, and regenerative transformation as intrinsic values.

This model outlines a synthetic strategy guided by three core capacities.

*Self-Dismantling.* Artificial systems are equipped to autonomously deconstruct themselves when specific environmental or functional thresholds are met.

*Modular Reuse.* Components are designed for recombination and use in new systemic roles or artifacts.

*Circular Integration.* Discarded materials and subsystems feed into ecological or artificial regenerative loops.

Rather than producing indefinitely persistent systems, techno-apoptotic design emphasizes temporality and reintegration. In this sense, it fully enacts the Gaian paradigm's emphasis on systemic viability, ecological resonance, and regenerative ethics.

We believe that robotics provides a particularly fertile terrain for this approach. As we have argued, contemporary robotics already functions as a conceptual and material frontier for testing sustainability-oriented synthetic models. Through the techno-apoptosis framework we intend to turn this potential into a programmatic design logic that bridges epistemological exploration and ecological responsibility.

### 5.2 Techno-Apoptosis in Gaian Synthetic Practice

Based on biological and systemic insights into apoptosis, Fleres [11] introduced the concept of techno-apoptosis. This refers to the design of artificial systems capable of self-dismantling and component recycling, and defines the intentional incorporation of finite lifecycles and modular renewal into robotic systems.

This groundwork finds a particularly compelling case of "implicit" techno-apoptosis in the history of Sony's AIBO robotic pet. As well-documented, after this robot's discontinuation in 2006 and the definitive end of official support in 2014, a distributed ecosystem spontaneously emerged, in which users, former Sony engineers, and informal communities began organizing repair, parts reuse, and symbolic rituals. Notably, A-Fun, a company dedicated to repairing AIBOs with salvaged components from other units, was founded. When a robot was irreparable, it was declared "dead" and dismantled, and its parts were reused in other AIBOs.

As discussed in previous works [7, 11–13], this emergent ecology enacted key features of techno-apoptosis: distributed network of agents coordinating disassembly and reuse; component compatibility and modularity enabled systems to "survive" through exchange; programmed disassembly realized *de facto* through repair and systematic redistribution. This example illustrates techno-apoptosis not only as a design hypothesis, but as a concrete phenomenon embedded in sociotechnical dynamics. In this way, it positions techno-apoptosis as a framework for experimental development, aiming at integrating robotic products into circular ecologies of use and reuse.

The specific contribution of this paper, building on this groundwork, is to develop the techno-apoptotic framework within the synthetic approach as a coherent expression of its Gaian paradigm. Our goal is to establish a positive synthetic loop, in which the scientific understanding of natural apoptotic processes, expanded through their multiple enactment in novel synthetic systems, informs techno-apoptosis as a concrete sustainability strategy for robotics.

#### 5.2.1 Techno-Apoptotic Scenarios in Electromechanical and Chemical Systems

To clarify how techno-apoptosis could operate within Gaian synthetic research, let us imagine an urban ecosystem—spanning interconnected homes, streets, and service infrastructures—populated by autonomous service robots. Each unit monitors its operational condition—battery life, sensor precision, energy consumption, and environmental footprint—while exchanging information within a distributed network. When degradation thresholds are reached, neighbouring robots initiate a coordinated dismantling protocol. Components such as sensors, microcontrollers, and batteries are catalogued, reconditioned, and redistributed through the network to sustain collective functionality. Through this distributed self-dismantling, the robotic ecology maintains resilience and circularity: *decay becomes a function of regeneration.* This scenario exemplifies how techno-apoptosis can work as a collective design principle, where individual disassembly enhances systemic sustainability. The process echoes the systemic logic of biological apoptosis: selective elimination that supports the viability of the whole. It also expands the scope of self-repair and adaptation by incorporating end-of-life transformation as a *designed* phase of robotic life, embedded within ecological and energetic cycles.

A parallel wetware case can be imagined in synthetic biology. Programmable protocells—expanding on early proposals for self-demolishing synthetic cells in nanomedicine [16]—are designed to perform catalytic remediation in polluted aquatic environments—detecting and neutralizing specific contaminants such as nitrates or hydrocarbons—and then initiate programmed self-demolition. Their biocompatible remnants are reabsorbed as nutrients by microbial communities or serve as molecular precursors for the spontaneous self-assembly of new protocells in the surrounding medium. This *wetware techno-apoptosis* illustrates how regenerative self-dissolution can operate at molecular scale, turning programmed decay into ecological contribution. In both robotic and biochemical domains, the same epistemic principle applies: sustainability emerges not from indefinite persistence but from the capacity to transform and reintegrate.

### 5.2.2 Advancing Techno-Apoptosis: A Gaian Synthetic Cycle

To move from the conceptual scenarios outlined above to a systematic design logic aligned with the Gaian paradigm, we propose a methodological cycle that integrates techno-apoptosis into the synthetic approach as both a theoretical and experimental framework. Its purpose is not to implement a specific robotic technology, but to operationalize the epistemic principle of *understanding by building within finitude*—that is, designing artificial systems that can sustain themselves through transformation and reintegration. (1) *Constructive Modelling.* Robotic or wetware systems are designed to instantiate hypotheses derived from natural apoptotic dynamics. Their architecture embodies key techno-apoptotic principles such as modularity, regenerative transformation, and programmed decay, enabling the experimental investigation of how synthetic systems can enact self-dismantling and material recycling as functional behaviours. (2) *Ecological Embedding.* These systems are situated within hybrid environments—technological, social, or ecological—where their interactions with local constraints generate emergent responses. The goal is not optimization, but adaptive integration: identifying how synthetic agents can become active participants in the metabolic and energetic flows that sustain their contexts. (3) *Comparative Exploration and Feedback.* The behaviour of techno-apoptotic systems is analysed comparatively with biological and social models of apoptosis. Divergences are treated not as failures but as diagnostic signals, exposing theoretical limitations and guiding new iterations of modelling. Each cycle thus yields dual feedback: it deepens scientific understanding of apoptotic principles and refines the design of sustainable artificial systems.

This Gaian synthetic cycle establishes techno-apoptosis as a *performative Gaian synthetic epistemology*: a way of building that generates knowledge through embedded experimentation and, at the same time, models an ethics of sustainability. It turns dismantling into a generative act—an exploration of how technology inspired by natural processes can contribute to planetary renewal.

## 6 Guidelines for a Gaian Robotics

Drawing on the epistemological, methodological, and theoretical developments articulated in this paper, we propose design guidelines rooted in the Gaian synthetic approach, using techno-apoptosis as a concrete operational instance.

*1. Circular and Participatory Design.* Robotic systems should be conceived for their entire lifecycle, integrating modularity, repairability, and disassembly from the earliest design stages. Components must be reusable and interoperable across different devices, supporting *programmed dismantling* and *systemic regeneration.* This circular architecture should extend beyond technical design to encompass participatory maintenance and repair cultures, in which human users, communities, and robotic agents co-manage the adaptive evolution of their shared ecosystems.

*2. Systemic Responsibility and Regenerative Integration.* Design and evaluation should shift from performance metrics to systemic contribution. Robotic systems must be embedded within networks of material and informational exchange that promote energy balance, resource circulation, and ecological resonance. Programmed device-decay thus becomes a deliberate design feature, clearly distinct from planned obsolescence, since its aim is not market turnover but systemic renewal—the reintegration of components and materials into broader techno-ecological cycles. These guidelines mark a first step in the transition from linear, efficiency-driven robotics toward a regenerative paradigm in which decay and renewal are intertwined. In the Gaian synthetic perspective, to sustain is to transform: technological life becomes sustainable precisely through its capacity to participate in the circular dynamics that sustain life itself.

## 7 Concluding Summary and Future Directions

The synthetic approach, grounded in the principle of *understanding by building*, can evolve into a methodological, technological, and ethical engagement with ecological challenges. Promoting this *Gaian turn*, we have outlined a vision of scientific and engineering practice that builds not only to understand, but to sustain. Artificial systems are here conceived as participants in planetary self-regulation—agents that can model, embody, and contribute to the circular dynamics of life.

Within this framework, *techno-apoptosis* has been introduced as a generative concept that translates the logic of biological and social self-regulation into synthetic design. Far from a metaphor, it offers a concrete strategy for integrating programmed dismantling, material reuse, and regenerative transformation into the architecture of robotic and wetware systems. Through this principle, sustainability is no longer an external constraint, but an intrinsic function of artificial construction.

This Gaian reconfiguration of the synthetic approach opens an interdisciplinary research trajectory that bridges philosophy, robotics, synthetic biology, and complex systems science. Future work will need to explore techno-apoptotic design not only as a proof of concept but as a *synthetic ethics* [5]—a mode of inquiry and creation capable of reimagining the relationship between technological and biological life.

In this perspective, the purpose of artificial construction is redefined: not the infinite expansion of control, but the conscious design of participation in planetary self-regulation.

**Acknowledgments.** This research has been developed within the PRIN 2022 project *Org(SB-EAI)* (n. 20222HHXAX, CUP F53D23004560006), funded by the Italian Ministry of University and Research under the European Union's NextGeneration EU, Mission 4, Component 1. The authors thank P. Stano and A. Roli for the fruitful scientific discussions on these topics within the ongoing collaboration in the Org (SB-EAI) project, with particular reference to WP3 and WP4.

## References

1. Brambilla, M., Ferrante, E., Birattari, M., Dorigo, M.: Swarm robotics: a review from the swarm engineering perspective. Swarm Intell. **7**, 1–41 (2013)
2. Capra, F., Luisi, P.L.: The Systems View of Life. Cambridge University Press (2014)
3. Cavalli, D., Chang, M., Milne, A., Morgan, W.: Xenophylum. Antikythera (2025, forthcoming). https://doi.org/10.1162/ANTI.5CZL
4. Cordeschi, R.: The Discovery of the Artificial: Springer, Dordrecht (2002)
5. Damiano, L.: Homes as human–robot ecologies: an epistemological inquiry on the "domestication" of robots. In: The Home in the Digital Age, pp. 80–102. Routledge (2021)
6. Damiano, L., Fleres, A., Roli, A., Stano, P.: Wetware network-based AI: a chemical approach to embodied cognition for robotics and artificial intelligence. Front. Robot. AI **12**, 1694338 (2026)
7. Damiano, L., & Fleres, A.: When Robots Are Abandoned: Post-discontinuation Sustainability as an Epistemological Problem (Preprint), Zenodo (2026)
8. Damiano, L., Hiolle, A., Cañamero, L.: Grounding synthetic knowledge. In: ECAL 2011 MIT Press (2011)
9. Damiano, L., Stano, P.: Explorative synthetic biology. Artif. Life**29**(3), 367–387 (2023)
10. Doncieux, S., Bredeche, N., Mouret, J.-B., Eiben, A.E.: Evolutionary robotics: what, why, and where to. Front. Robot. AI **2**, 4 (2015)
11. Fleres, A.: Robotica sociale e sostenibilità. Impalcature teoriche della robosfera. Ph.D. thesis, IULM University, Milan (2025). Under embargo
12. Fleres, A., Damiano, L.: AIBO and the emergence of the robosphere. In: Proceedings of CIFMA 2025 (2025). In press
13. Fleres, A., Damiano, L.: From AIBO to robosphere. organizational interdependencies. Front. Robot. AI **12**, 1716801 (2025b)
14. Ihle, K.E., de Guzman, L.I., Danka, R.G.: Social apoptosis in Varroa mite resistant western honey bees (Apis mellifera). J. Insect Sci. **22**(1), 13–14 (2022)
15. Laschi, C., Mazzolai, B., Cianchetti, M.: Soft robotics. Sci. Robot. **1**(1) (2016)
16. Leduc, P.R., et al.: Towards an in vivo biologically inspired nanofactory. Nat. Nanotechnol. **2**(1), 3–7 (2007)
17. Lemmens, P. Hui, Y.: Reframing the technosphere. Krisis **2**, 26–41 (2017)
18. Lovelock, J.E.: Gaia: a New Look at Life on Earth. Oxford University Press (1979)
19. Lovelock, J.E.: Novacene: The Coming Age of Hyperintelligence. The MIT Press (2019)
20. Lovelock, J.E., Margulis, L.: Atmospheric homeostasis by and for the biosphere: the Gaia hypothesis. Tellus **26**(1–2), 2–10 (1974)
21. Morin, E., Kern, A.B.: Homeland Earth. Hampton Press, Cresskill (1999)
22. Pfeifer, R., Scheier, C.: Understanding Intelligence. MIT Press, Cambridge (1999)
23. Pickering, A.: The Cybernetic Brain. University of Chicago Press, Chicago (2010)
24. Ross, T.: Machines that think. Sci. Am. **148**(4), 206–208 (1933)
25. Sloterdijk, P., Heinrichs, H.J.: Neither sun nor death. Semiotext(e) (2011)
26. Stano, P.: Minimal cells. Biotechnol. J. **6**, 850–859 (2011)

27. Stano, P.: Chemical systems for wetware artificial life: selected perspectives in synthetic cell research. Int. J. Mol. Sci. **24**, 14138 (2023)
28. Steels, L., Brooks, R. (eds.): The AL Route to AI. Routledge, London (2018)
29. van der Hout, S.: The homeotechnological turn: Sloterdijk's response to the ecological crisis. Environ. Values **23**(4), 423–442 (2014)

# Impact of the COVID-19 Outbreak on the Italian Twitter Vaccination Debate: A Network-Based Analysis

Veronica Lachi[1(✉)], Giovanna Maria Dimitri[2], Alessandro Di Stefano[3], Pietro Liò[4], Monica Bianchini[2], and Chiara Mocenni[2]

[1] UiT, The Arctic University of Norway, Tromsø, Norway
veronica.lachi@uit.no
[2] Department of Information Engineering and Mathematics, University of Siena, Siena, Italy
[3] School of Computing, Engineering and Digital Technologies, Teesside University, Middlesbrough, UK
[4] Department of Computer Science and Technology, University of Cambridge, Cambridge, UK

**Abstract.** This paper investigates the impact of the COVID-19 outbreak on the Italian Twitter vaccination debate through a network-based analysis. By collecting over 9 million Italian-language tweets from January 2019 to May 2022, we constructed monthly retweet networks to study the evolution of user interactions and polarization. Our results show a dramatic increase in activity and network centralization after the pandemic onset, with the debate becoming more polarized and dominated by a few influential hubs. We identified and characterized core ProVax and NoVax communities, highlighting their differences in engagement, visibility, and influence. The publicly released TwitterVax dataset supports further research on vaccine-related discourse in social media.

**Keywords:** Twitter · COVID-19 · vaccination · Pro/NoVax community detection · social networks

## 1 Introduction

Vaccines have historically played a crucial role in reducing mortality from infectious diseases [20]. Yet, the phenomenon of *vaccine hesitancy*—the delay or refusal of vaccination despite availability—is increasingly widespread, fueled by online misinformation and amplified through social media [42,43]. This multifaceted attitude is shaped by socio-economic, psychological and cultural factors [10,13,24].

The COVID-19 pandemic has brought renewed attention to vaccine hesitancy. Although vaccines proved more effective than measures such as lockdowns [1,27], a significant portion of the population resisted inoculation, often influenced by distrust, conspiracy theories, and polarizing online discourse [11,16,32]. Social media, and Twitter in particular [2], became central to the dissemination of both scientific information and anti-vaccine sentiment.


M. Bianchini et al. (Eds.): WIVACE 2025, CCIS 3001, pp. 171–185, 2027.
https://doi.org/10.1007/978-3-032-33185-4_13

In this paper, we analyze the Italian Twitter vaccination debate from January 2019 to May 2022 using a network-based approach. We show how the pandemic reshaped the structure of the debate, increasing polarization and concentrating influence in a few hubs. We identify ProVax and NoVax communities, study their temporal dynamics and productivity, and release the `TwitterVax` dataset[1] to support future research. The rest of the paper is organized as follows: Sect. 2 reviews related literature; Sect. 3 describes our methodology; Sect. 4 presents the results; and Sect. 5 concludes the study.

## 2 Related Works

Several studies have examined the vaccination debate on social media using network-based approaches. Analyses of Twitter data have shown that user opinions can influence vaccination decisions and that communities of vaccine supporters and opponents display distinct structural patterns, with skeptics often forming more cohesive clusters [8,15,31,34]. While prior work has mostly focused on the static structure of these communities, fewer studies have explored the impact of the COVID-19 pandemic on vaccine discourse. Notably, [12] reported a sharp increase in NoVax activity following the outbreak, and [36] showed sentiment shifts in early COVID-19 tweets. Regarding Italy, limited research exists. Some works analyzed vaccine-related tweets over short periods [9,23], or within multilingual global datasets [29]. A broader study in [17] confirmed the persistence of echo chambers during the pandemic, but lacked a fine-grained structural analysis. Our work contributes to this line of research by analyzing the Italian Twitter vaccine debate over three years. We focus on the structural evolution of the ProVax and NoVax communities, quantifying their relative size, activity, and core user composition over time.

## 3 Materials and Methods

Twitter, a widely recognized social networking platform, had its data publicly accessible via its application programming interface (API) until February 2023. In this paper, we use a network–based approach to study the evolution of Italian public opinion on vaccines over time. Transforming the unstructured information available from Twitter into graph data has allowed us not only to employ common metrics used on networks to estimate their topological features but also to apply some powerful algorithms for community detection.

### 3.1 Data Collection

We built our dataset using the Twitter API v2, specifically the `tweets/search/all` endpoint[2], which grants access to the full historical archive of tweets. We

[1] Available at https://github.com/veronicalachi/TwitterVax.
[2] https://developer.twitter.com/en/docs/twitter-api/tweets/search/api-reference/get-tweets-search-all.

retrieved all Italian-language tweets related to vaccination, posted between January 1, 2019, and May 31, 2022, based on a set of twenty vaccine-related keywords: *vaccino, vaccini, vaccinazione, vaccinazioni, vaccinare, vaccinarsi, vaccinatevi, vacciniamoci, vaccinando, vaccinale, vaccinali, vaccinati, vaccinate, vaccinata, vaccinato, va@@ino, va..ino, vaxino, vaxxino, #iomiovaccino.* Some of these terms, such as *vaxxino*, are domain-specific and frequently used by both ProVax and NoVax communities. Unlike previous studies (e.g., [17]), we limited the keyword selection to Italian terms and filtered tweets to retain only those written in Italian. The resulting `TwitterVax` dataset comprises 9,068,389 tweets authored by 300,653 unique users. To capture temporal dynamics, we divided the data into 41 monthly sub-periods. This one-month granularity was heuristically chosen as it best preserved the evolution of community structures over time. It is important to note that Twitter's Search API may omit tweets deleted by users or removed due to policy violations. As highlighted by [38], up to 72% of tweets collected via the Streaming API at posting time were no longer available through the Historical Search API. This emphasizes the transient nature of Twitter content and the potential underestimation of controversial or policy-violating material in retrospective analyses.

### 3.2 Network Construction and Indicators

For each month in the dataset, we built a retweet network where nodes represent active users and undirected edges connect users who retweeted each other at least once. This choice addresses a limitation in [17], where edges required at least two interactions, excluding over half the users. Since retweeting often signals endorsement [5,21,33], the resulting graphs can be interpreted as *agreement networks*, reflecting shared opinions.

To focus on the main discussion dynamics, we retained only the largest connected component of each graph. This allowed us to capture the temporal evolution of the debate while reducing noise. The following metrics were computed to characterize network structure:

- **Density**—The ratio between observed and possible edges, indicating how densely users are connected [44].
- **Average clustering coefficient**—Measures the tendency of a node's neighbors to form triangles, computed as:

$$c_v = \frac{2T(v)}{\deg(v)(\deg(v)-1)}$$

  where $T(v)$ is the number of triangles through $v$ and $\deg(v)$ its degree [44,45].
- **s–metric**—A structural measure of scale-freeness, defined as:

$$s(\mathcal{G}) = \sum_{(v,u)\in\mathcal{E}} \deg(v)\cdot\deg(u)$$

  A high $s$-metric indicates the presence of hubs typical of scale-free networks [4,30].

### 3.3 Community Detection and User Labeling

Given the polarized nature of the vaccination debate [41], we assume that each user aligns with either the ProVax or NoVax stance. However, classifying users based on tweet content is unfeasible due to the large dataset size and the intrinsic limitations of NLP techniques on short, semantically complex tweets [35,37]. Therefore, we infer user leanings through network structure.

We identify communities of users using the METIS graph partitioning algorithm [26], which is well suited to retweet networks [15,21]. METIS offers high-quality partitions and scalability, and allows for pre-specification of the number of communities—two in our case.

The algorithm operates in two phases:

1. *Coarsening*: the input graph $\mathcal{G}_0$ is compressed into successively smaller graphs $\mathcal{G}_1, \ldots, \mathcal{G}_k$, enabling fast partitioning at the coarsest level.
2. *Refinement*: the coarse partition $\mathcal{P}_k$ is iteratively projected and refined back onto $\mathcal{P}_0$.

We ran METIS 100 times with two partitions. Each user's leaning is defined as the average over their partition assignments, yielding a continuous score in $[0, 1]$. To tune the partition size hyperparameter, we maximized the number of users with scores within the 95% confidence interval.

We manually labeled the top 10% of users with extreme leaning scores. Remaining users were assigned to the same community as the closest extreme user. This approach enabled temporal tracking of the ProVax and NoVax communities and their polarization.

Following [14], we computed a polarization score $S_t$ at each time $t$ as:

$$S_t = \frac{(E_n + E_p - E_o)}{(E_n + E_p + E_o)} \tag{1}$$

where $E_n$ and $E_p$ denote the intra-community edge densities for NoVax and ProVax users, respectively, and $E_o$ the inter-community edge density. A value of $S_t$ close to 1 indicates strong polarization, while values near $-1$ reflect more mixed interaction patterns.

### 3.4 Vaccination Debate Analysis Through the Global Multiplexity Matrix

To analyze the temporal stability of user communities, we constructed a *global multiplexity matrix* $M$, a tool commonly used in dynamic network analysis [3,25]. In our formulation, both rows and columns of $M$ correspond to all users who participated in the vaccine debate at any time, and each entry $M(i, j)$ indicates how many times users $i$ and $j$ were assigned to the same community across the 41 monthly networks. Although our monthly graphs do not form a proper multiplex network—defined as a collection of layers with identical node sets [6,7,18]—this aggregated approach allows us to detect persistent associations. We focused on

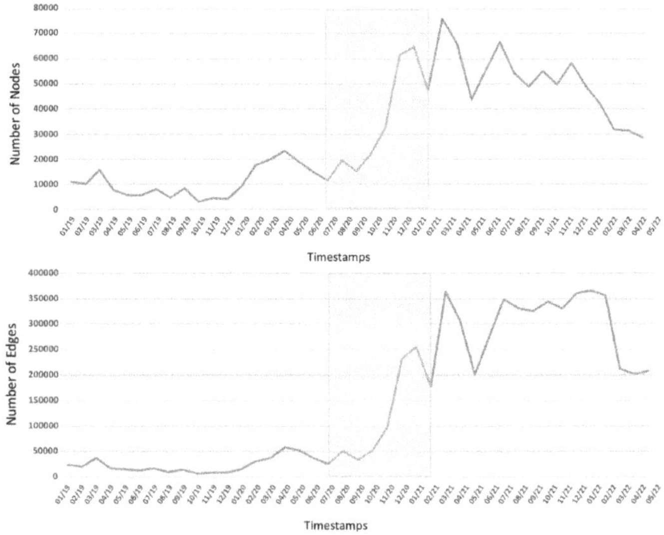


**Fig. 1.** Temporal evolution of the number of nodes (left) and edges (right) in the monthly retweet networks. The sharp increase starting around October 2020 aligns with the second COVID-19 wave in Italy.

identifying users who were consistently part of the same community throughout the entire period, revealing the most stable and committed ProVax and NoVax cores. We further investigated the characteristics of these core groups by checking whether users were *verified* (i.e., marked with Twitter's blue checkmark to confirm identity authenticity). Verified users often include public figures or institutions (e.g., journalists, politicians, or media outlets) and play a crucial role in information diffusion. Thus, we also measured their follower counts and network connectivity to assess their influence and reach within the debate.

# 4 Results and Discussion

## 4.1 Temporal Evolution of the Network Structure

We constructed 41 monthly retweet networks, where each node represents an active user and each undirected edge a retweet interaction. In every month, the largest connected component includes over 90% of the nodes. As shown in Fig. 1, the number of users and interactions steadily increases from early 2019 to late 2021, peaking around October 2020—coinciding with the second wave of COVID-19 in Italy and the anticipation of vaccine rollout. The discussion volume remains elevated throughout 2021, then gradually declines from April 2022 onward, likely due to pandemic containment and waning public interest.

To better understand how the structure of the debate evolved, we computed four network metrics over time: density, average clustering coefficient, average node degree, and the $s$–metric. As shown in Fig. 2, density and clustering coefficient decrease over time, indicating that as participation increases, users form fewer local clusters. Conversely, average degree and $s$–metric increase, highlighting the emergence of influential hubs—users with very high connectivitysuggesting a shift toward a scale-free structure [46]. Initially, the debate was more evenly distributed, but over time it became increasingly centralized, dominated by a small number of highly retweeted accounts. This transformation is further confirmed by the power-law distribution of node degrees (Fig. 3). The trend partially reverts in early 2022, as the debate shrinks and the network regains structural similarities to the pre-pandemic phase. Monitoring these dynamics is essential, as the network structure directly influences how (mis)information spreads [22]. While sparsity can facilitate rapid diffusion, it also amplifies the risk of disinformation, especially if malicious actors occupy central positions in the network. Lower clustering further hampers efforts to contain false narratives, as information bypasses traditional verification bottlenecks.

### 4.2 Temporal Evolution of the Community Composition and Network Polarization

As discussed in Sect. 3.3, we applied the METIS algorithm 100 times to each of the 41 monthly retweet networks, resulting in community assignments from which we derived continuous leaning scores. Figure 4 shows the heatmap of these scores over time. A growing concentration of users toward extreme values (0 or 1) suggests increasingly clear polarization and improved separability of the two communities. An exception is visible between August 2020 and February 2021, a period marked by a regionalized lockdown system and the early stages of the vaccination campaign. This likely introduced uncertainty and temporarily softened the distinction between user groups.

We manually labeled users with the most extreme scores to assign ProVax (blue) and NoVax (orange) affiliations, which were then propagated to the rest of the network. An example partition is shown in Fig. 5.

Figure 6 shows the relative size of the two communities across time. Until early 2021, ProVax users generally outnumbered NoVax ones. From April 2021 onward, however, NoVax users became progressively more numerous, surpassing ProVax users by January 2022.

Activity levels followed a similar trend. As shown in Fig. 7, NoVax users consistently produced more tweets than ProVax users, except during the initial five months of the vaccination campaign. From April 2021 onward, NoVax productivity rose, while ProVax participation declined.

Peaks in ProVax activity coincide with major milestones of the national vaccination campaign (e.g., rollout to medical personnel and first three doses), while NoVax activity peaks correspond to the green pass mandate (August 2021) and the end of the outdoor mask requirement (February 2022). This divergence suggests differing motivational drivers: civic engagement for ProVax users versus personal liberty concerns for NoVax users. These patterns align with recent studies showing that vaccine hesitancy is correlated with lower levels of social

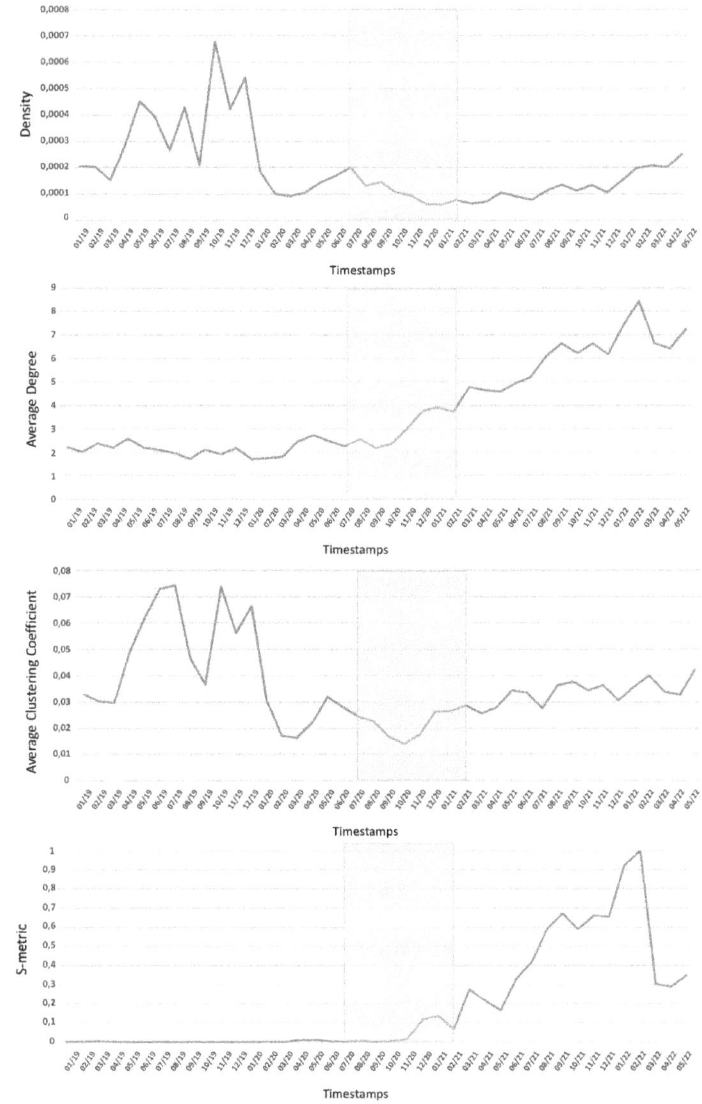


**Fig. 2.** Evolution of network metrics over time. In all the plots, the $x$ axis shows the month numbers, while the $y$ axis describes the values of the four network measures considered (density, average clustering coefficient, average degree and s–metric). The yellow box indicates the transition period, reported also in Fig. 1, approximately ranging from July 2020 to February 2021. (Color figure online)

trust and weaker adherence to collective norms [19,28,40]. Finally, we computed the polarization score $S_t$ defined in Eq. (1) for each month. As shown in Fig. 8, polarization increased steadily over time, with a transient dip around December 2020 ("Vax-day"), suggesting a brief phase of unified sentiment. However, the

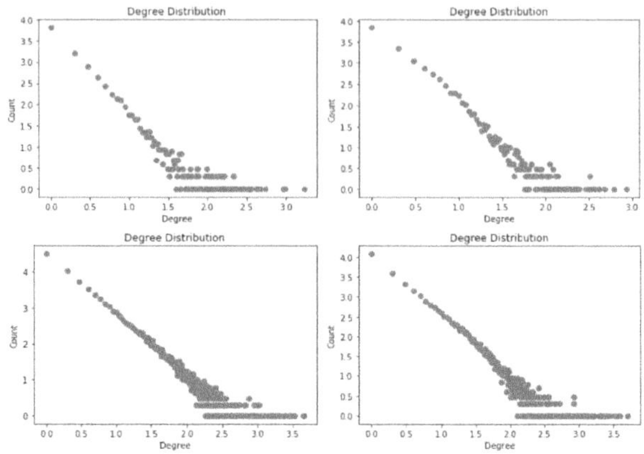


**Fig. 3.** Degree distribution of the retweet networks at months 1, 21, 31, 41. All the networks are scale–free.

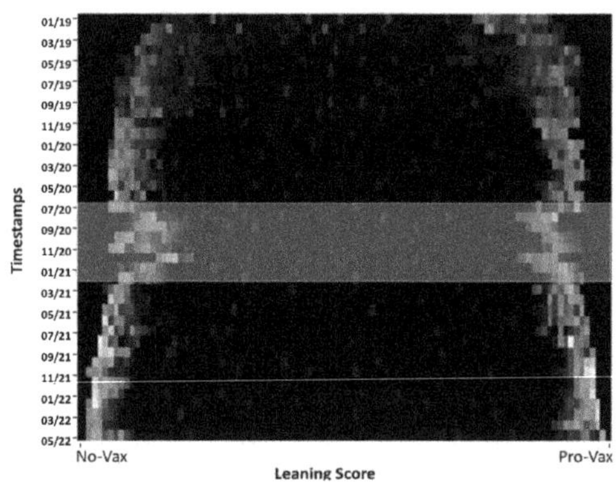


**Fig. 4.** Heatmap of leaning scores over time. Lighter colors indicate a higher proportion of users with a given leaning score.

overall trend supports previous findings that COVID-19 amplified, rather than mitigated, existing echo chambers in vaccine discourse [17].

### 4.3 Possible Strategies to Slow down Vaccine Misinformation

By identifying communities over time, we can develop effective strategies to curb the spread of vaccine misinformation and limit the growth of the NoVax community. One promising approach is to pinpoint nodes with a central role in the NoVax community and disrupt their influence through node attacks, such as account blocking. To determine node importance, we employed the betweenness centrality metric for NoVax nodes. Analyzing the betweenness of NoVax nodes enabled us to examine the pathways of misinformation spreading. Specifically, we analyzed the subgraph induced by the NoVax nodes at each time step and cal-

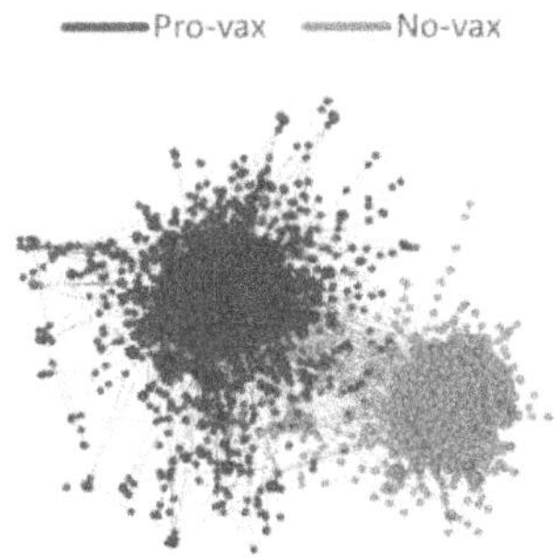


Fig. 5. Example of community-labeled retweet network (Month 1). (Color figure online)

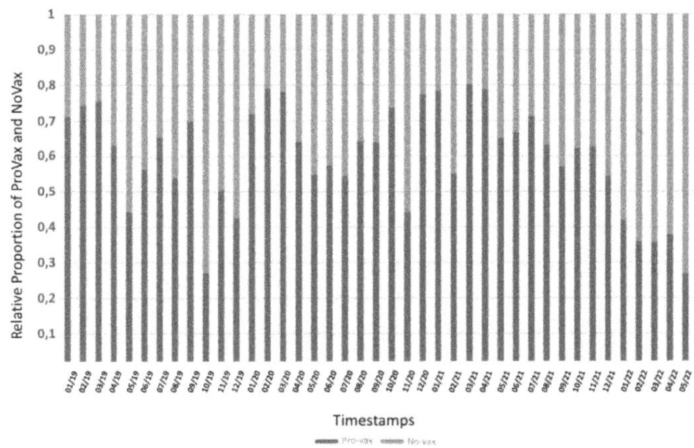


Fig. 6. Relative proportion of NoVax (orange) and ProVax (blue) users over time. (Color figure online)

culated the distribution of betweenness among the nodes in the subgraph. Our analysis revealed that all the distributions followed a power–law trend, indicating that only few nodes had high betweenness at each time step. We then determined the number of nodes with extremely high betweenness by calculating the percentage of nodes outside the 95th percentile of the distribution for each time step (Fig. 9). Although the number of nodes with high betweenness increased over time, they constituted a small portion of the total nodes. Therefore, account blocking could be an effective strategy to hinder their ability to spread misinformation. Other node attacks, based on metrics such as degree and closeness, could also fragment the NoVax community.

### 4.4 Characterization of the Core NoVax/ProVax Users

The global multiplexity matrix $M$, of dimension 300,653×300,653, is found to be sparse. Each entry $M(i,j)$ is an integer between 0 and 41 representing the

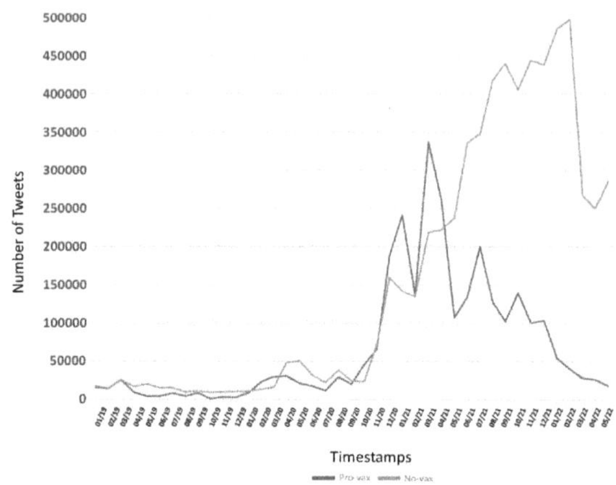


**Fig. 7.** Monthly tweet volume for NoVax (orange) and ProVax (blue) users. (Color figure online)

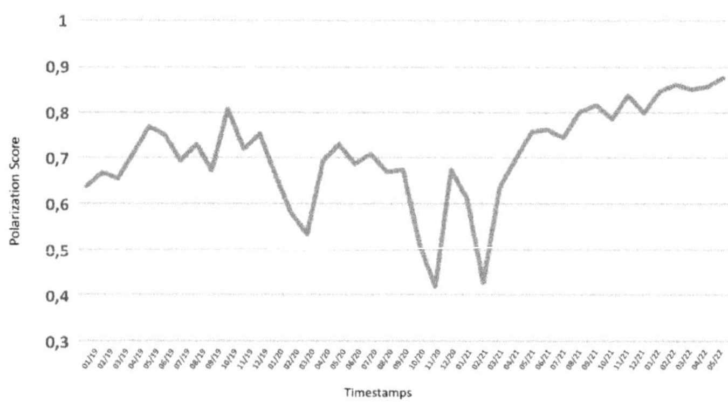


**Fig. 8.** Temporal evolution of the polarization score $S_t$.

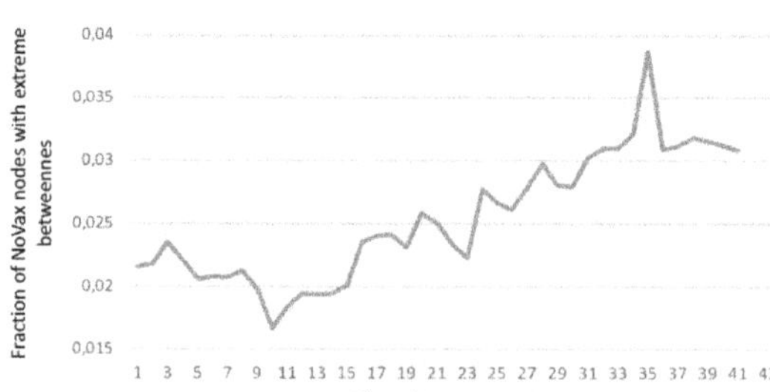


**Fig. 9.** Fraction of NoVax users characterized by a betweenness centrality outside the 95th percentile.

number of months in which the two users $i$ and $j$ belong to the same community. First, we extracted all pairs of users who have been in the same community throughout the period, i.e. pairs of users whose relative matrix entry equals 41. Then, we used the extracted information to construct a new network connecting only such selected users. The resulting graph has seven connected components, whose dimensions are shown in Table 1.

**Table 1.** Dimension and composition of the seven connected components composed by users who belong to the same community for forty–one months; the percentage is calculated as the relative number of times that the nodes inside the connected components represent NoVax/ProVax users

| # Nodes in Connected Component | NoVax | ProVax |
|---|---|---|
| 183 | **100%** | 0% |
| 10 | 2.4% | **97.6%** |
| 8 | **95.1%** | 4.9% |
| 4 | 4.8% | **95.2%** |
| 4 | 4.9% | **95.1%** |
| 2 | 5.1% | **94.9%** |
| 2 | **97.5%** | 2.5% |

It is worth noting that each component is fully connected because requiring that nodes belong to the same community forty–one times—which is the maximum possible—is a transitive membership relation. As we can see in Table 1, the largest connected component is totally composed of users that have always been NoVax, whereas ProVax convinced users populate the second largest component. The other five connected components are composed by few users that have changed their mind about vaccines at least one time during the forty–one months. However, it is also possible that some convinced and active users have not tweeted during some periods. Considering this, we decided to relax the constraint by requiring that nodes belong to the same community just forty times. In this case, the graph has four connected components—no more fully–connected —, with the largest one composed by the core NoVax users (Table 2).

**Table 2.** Dimension and composition of the four connected components composed by users who belong to the same community for forty months

| # Nodes in Connected Component | NoVax | ProVax |
|---|---|---|
| 312 | **96.5%** | 3.5% |
| 60 | 3.2% | **96.8%** |
| 2 | **95.1%** | 4.9% |
| 2 | **95%** | 5% |

Decreasing the number of periods to thirty–nine, the two core NoVax and ProVax sets became larger and more stable: the graph at this point contains only two connected components, whose dimension and composition are shown in Table 3. Based on Table 3, we can also conclude that the NoVax supporters are more numerous and stable over time, whereas the core set of ProVax users is significantly smaller. Anyway, the most convinced and active ProVax users have 100200 followers on average, while NoVaxes have generally less followers (44067), even if they are more constant and more present in the debate. Furthermore, none of the users who make up the NoVax backbone are verified users, while 15.85% of vaccine advocates are.

Another noteworthy observation pertains to the structure of the core NoVax and ProVax users. NoVax users exhibit a significantly more homogeneous structure compared to the ProVax users. This characteristic may be closely tied to their efficacy in disseminating false information. As highlighted in [39], the act of sharing false information, even if subsequently debunked, tends to correlate with individuals either postponing vaccination or outright rejecting it.

**Table 3.** Dimension and composition of the two connected components composed by users who belong to the same community for thirty–nine months

| # Nodes in Connected Component | NoVax | ProVax |
|---|---|---|
| 405 | **95.6%** | 4.4% |
| 97 | 3.3% | **96.7%** |

## 5 Conclusion

This study analyzed the evolution of the Italian vaccination debate on Twitter from January 2019 to May 2022 using the newly collected TwitterVax dataset. Our network-based approach revealed that the COVID-19 outbreak significantly amplified and polarized the discussion, concentrating interactions around a few influential hubs. Although this centralization faded after early 2022, polarization between ProVax and NoVax users steadily increased. We found that NoVax users, though fewer, were more active, while ProVax users were more often verified and had broader reach. A multiplex analysis identified stable core communities, highlighting key differences in engagement and influence.

## References

1. Alfano, V., Ercolano, S.: The efficacy of lockdown against covid-19: a cross-country panel analysis. Appl. Health Econ. Health Policy **18**, 509–517 (2020)
2. Auxier, B., Anderson, M.: Social media use in 2021. Pew Res. Center **1**, 1–4 (2021)

3. Azevedo, T., Dimitri, G.M., Lió, P., Gamazon, E.R.: Multilayer modelling of the human transcriptome and biological mechanisms of complex diseases and traits. NPJ Syst. Biol. Appl. **7**(1), 1–13 (2021)
4. Barabási, A.-L., Albert, R.: Emergence of scaling in random networks. Science **286**(5439), 509–512 (1999)
5. Barberá, P., Jost, J.T., Nagler, J., Tucker, J.A., Bonneau, R.: Tweeting from left to right: is online political communication more than an echo chamber? Psychol. Sci. **26**(10), 1531–1542 (2015)
6. Battiston, F., Nicosia, V., Latora, V.: Structural measures for multiplex networks. Phys. Rev. E **89**(3), 032804 (2014)
7. Battiston, F., Nicosia, V., Latora, V.: The new challenges of multiplex networks: measures and models. Eur. Phys. J. Spec. Top. **226**(3), 401–416 (2017)
8. Bello-Orgaz, G., Hernandez-Castro, J., Camacho, D.: Detecting discussion communities on vaccination in twitter. Futur. Gener. Comput. Syst. **66**, 125–136 (2017)
9. Bellodi, E., Bertagnon, A., Gavanelli, M.: Comparing emotion and sentiment analysis tools on Italian anti-vaccination for covid-19 posts. In: Proceedings of the Sixth Workshop on Natural Language for Artificial Intelligence (NL4AI 2022) co-located with 21th International Conference of the Italian Association for Artificial Intelligence (AI* IA 2022) (2022)
10. Bertoncello, C., et al.: Socioeconomic determinants in vaccine hesitancy and vaccine refusal in Italy. Vaccines **8**(2), 276 (2020)
11. Betsch, C., et al.: Opportunities and challenges of web 2.0 for vaccination decisions. Vaccine **30**(25), 3727–3733 (2012)
12. Bonnevie, E., Gallegos-Jeffrey, A., Goldbarg, J., Byrd, B., Smyser, J.: Quantifying the rise of vaccine opposition on twitter during the covid-19 pandemic. J. Commun. Healthc. **14**(1), 12–19 (2021)
13. Browne, M., Thomson, P., Rockloff, M.J., Pennycook, G.: Going against the herd: psychological and cultural factors underlying the 'vaccination confidence gap'. PLoS ONE **10**(9), e0132562 (2015)
14. Chen, T.H.Y., Salloum, A., Gronow, A., Ylä-Anttila, T., Kivelä, M.: Polarization of climate politics results from partisan sorting: evidence from finnish twittersphere. Glob. Environ. Chang. **71**, 102348 (2021)
15. Cossard, A., De Francisci, G., Morales, K.K., Mejova, Y., Paolotti, D., Starnini, M.: Falling into the echo chamber: the Italian vaccination debate on Twitter. In: Proceedings of the International AAAI Conference on Web and Social Media, vol. 14, pp. 130–140 (2020)
16. Cota, W., Ferreira, S.C., Pastor-Satorras, R., Starnini, M.: Quantifying echo chamber effects in information spreading over political communication networks. EPJ Data Sci. **8**(1), 1–13 (2019)
17. Crupi, G., Mejova, Y., Tizzani, M., Paolotti, D., Panisson, A.: Echoes through time: evolution of the Italian covid-19 vaccination debate. In: Proceedings of the International AAAI Conference on Web and Social Media, vol. 16, pp. 102–113 (2022)
18. Dimitri, G.M., et al.: A multiplex network approach for the analysis of intracranial pressure and heart rate data in traumatic brain injured patients. Appl. Netw. Sci. **2**(1), 1–12 (2017)
19. Faezi, N.A., et al.: Peoples' attitude toward covid-19 vaccine, acceptance, and social trust among African and middle east countries. Health Promot. Perspect. **11**(2), 171 (2021)

20. Centers for Disease Control, Prevention (CDC). Impact of vaccines universally recommended for children–united states, 1990-1998. MMWR Morbidity Mortality Weekly Report **48**(12), 243–248 (1999)
21. Garimella, K., De Francisci, G., Morales, A.G., Mathioudakis, M.: Quantifying controversy on social media. ACM Trans. Soc. Comput. **1**(1), 1–27 (2018)
22. Gleeson, J.P., O'Sullivan, K.P., Baños, R.A., Moreno, Y.: Effects of network structure, competition and memory time on social spreading phenomena. Phys. Rev. X **6**(2), 021019 (2016)
23. Gori, D., et al.: Mis-tweeting communication: a vaccine hesitancy analysis among twitter users in Italy. Acta Bio Medica Atenei Parmensis **92**(Suppl. 6) (2021)
24. Hornsey, M.J., Harris, E.A., Fielding, K.S.: The psychological roots of anti-vaccination attitudes: a 24-nation investigation. Health Psychol. **37**(4), 307 (2018)
25. Hristova, D., Rutherford, A., Anson, J., Luengo-Oroz, M., Mascolo, C.: The international postal network and other global flows as proxies for national wellbeing. PLoS ONE **11**(6), e0155976 (2016)
26. Karypis, G., Kumar, V.: A fast and high quality multilevel scheme for partitioning irregular graphs. SIAM J. Sci. Comput. **20**(1), 359–392 (1998)
27. Kim, J.H., Marks, F., Clemens, J.D.: Looking beyond covid-19 vaccine phase 3 trials. Nat. Med. **27**(2), 205–211 (2021)
28. Kreps, S.E., Kriner, D.L.: Resistance to covid-19 vaccination and the social contract: evidence from Italy. NPJ Vaccines **8**(1), 60 (2023)
29. Lenti, J., et al.: Global misinformation spillovers in the vaccination debate before and during the covid-19 pandemic: multilingual twitter study. JMIR Infodemiol. **3**, e44714 (2023)
30. Li, L., Alderson, D., Doyle, J.C., Willinger, W.: Towards a theory of scale-free graphs: definition, properties, and implications. Internet Math. **2**(4), 431–523 (2005)
31. Lutkenhaus, R.O., Jansz, J., Bouman, M.P.A.: Mapping the Dutch vaccination debate on twitter: identifying communities, narratives, and interactions. Vaccine X **1**, 100019 (2019)
32. Machingaidze, S., Wiysonge, C.S.: Understanding covid-19 vaccine hesitancy. Nat. Med. **27**(8), 1338–1339 (2021)
33. Metaxas, P., Mustafaraj, E., Wong, K., Zeng, L., O'Keefe, M., Finn, S.: What do retweets indicate? Results from user survey and meta-review of research. In: Proceedings of the International AAAI Conference on Web and Social Media, vol. 9, pp. 658–661 (2015)
34. Milani, E., Weitkamp, E., Webb, P.: The visual vaccine debate on twitter: a social network analysis. Media Commun. **8**(2), 364–375 (2020)
35. Müller, M., Salathé, M., Kummervold, P.E.: Covid-twitter-bert: a natural language processing model to analyse covid-19 content on twitter. arXiv preprint arXiv:2005.07503 (2020)
36. Naseem, U., Razzak, I., Khushi, M., Eklund, P.W., Kim, J.: COVIDSenti: a large-scale benchmark twitter data set for COVID-19 sentiment analysis. IEEE Trans. Comput. Soc. Syst. **8**, 1003–1015 (2021)
37. Nguyen, A., Longa, A., Luca, M., Kaul, J., Lopez, G.: Emotion analysis using multilayered networks for graphical representation of tweets. IEEE Access **10**, 99467–99478 (2022)
38. Pfeffer, J., Mooseder, A., Hammer, L., Stritzel, O., Garcia, D.: This sample seems to be good enough! assessing coverage and temporal reliability of twitter's academic API. arXiv preprint arXiv:2204.02290 (2022)

39. Pierri, F., et al.: Online misinformation is linked to early covid-19 vaccination hesitancy and refusal. Sci. Rep. **12**(1), 5966 (2022)
40. Roy, D.N., Biswas, M., Islam, E., Azam, M.S.: Potential factors influencing covid-19 vaccine acceptance and hesitancy: a systematic review. PLoS ONE **17**(3), e0265496 (2022)
41. Schmidt, A.L., Zollo, F., Scala, A., Betsch, C., Quattrociocchi, W.: Polarization of the vaccination debate on Facebook. Vaccine **36**(25), 3606–3612 (2018)
42. Schuster, M., Eskola, J., Duclos, P., et al.: Review of vaccine hesitancy: rationale, remit and methods. Vaccine **33**(34), 4157–4160 (2015)
43. Simas, C., Larson, H.J.: Overcoming vaccine hesitancy in low-income and middle-income regions. Nat. Rev. Dis. Primers. **7**, 41 (2021)
44. Van Steen, M.: Graph theory and complex networks. An introduction, p. 144 (2010)
45. Watts, D.J., Strogatz, S.H.: Collective dynamics of 'small-world' networks. Nature **393**(6684), 440–442 (1998)
46. Xiaoyan, W., Liu, Z.: How community structure influences epidemic spread in social networks. Physica A **387**(2–3), 623–630 (2008)

# Numerical Modeling of Surface Reactions in Chemoresistive Gas Sensors: A Computational Framework for Predictive Analysis

Ada Fort, Enza Panzardi(✉), Marco Mugnaini, and Valerio Vignoli

Department of Information Engineering and Mathematics, University of Siena, Via Roma 56, 53100 Siena, Italy
enza.panzardi@unisi.it

**Abstract.** Chemoresistive gas sensors based on metal oxides are widely used in environmental monitoring, industrial safety, and smart systems. Despite their commercial relevance, the underlying physico-chemical processes that govern their electrical response remain partially understood, limiting the development of robust, drift-resistant, and adaptive sensor systems. This paper investigates the potential of numerical modeling as a computational approach to describe and predict the dynamic behavior of these sensors under varying gas compositions and thermal conditions.

We present the surface-state–based modeling framework that integrates the kinetics of surface reactions with electronic conduction mechanisms, focusing on n-type metal oxide materials. By translating the complex interaction between gas species and the sensing material into a system of equations, the model enables simulation of the transient and steady-state conductance. This approach provides not only a compact description of sensor behavior but also a foundation for algorithmic learning and control strategies in complex environments. Our findings highlight how such numerical models can serve as interpretable algorithmic abstractions, bridging physics-based sensor modeling with data-driven inference in artificial intelligence systems.

**Keywords:** gas sensors · gas sensors response model · $SnO_2$ sensor · CO sensing

## 1 Introduction

In recent decades, significant research efforts have been dedicated to understanding the sensing mechanisms of conductometric metal oxide (MOX) gas sensors [1]. Particular attention has been given to materials such as $ZnO$, $SnO_2$, $TiO_2$, $In_2O_3$, and $WO_3$, which form the foundation of many commercial gas-sensing technologies. These materials are especially relevant due to their excellent sensitivity, selectivity, and stability when exposed to toxic gases commonly produced during combustion processes [2].

Combustion-related gases—such as carbon monoxide (CO), nitrogen oxides ($NO_x$), and carbon dioxide ($CO_2$)—have garnered increasing attention not only because of their


M. Bianchini et al. (Eds.): WIVACE 2025, CCIS 3001, pp. 186–196, 2027.
https://doi.org/10.1007/978-3-032-33185-4_14

immediate impact on human health but also due to their critical role in environmental pollution and climate change. CO and $NO_x$ are among the most prevalent air pollutants in urban and industrial environments, contributing to smog formation, acid rain, and respiratory diseases [3]. $CO_2$, while non-toxic at typical atmospheric levels, is the primary greenhouse gas emitted through anthropogenic activities and a major driver of global warming [4]. In this context, the development and deployment of effective gas-sensing technologies is not only desirable, but essential to support global efforts aimed at climate mitigation and environmental protection.

However, environmental monitoring is only one among the many domains in which MOX gas sensors have demonstrated their relevance. Thanks to their compactness, low cost, and compatibility with microelectronic platforms, MOX sensors are increasingly used in other critical applications, including clinical diagnostics (e.g., breath analysis for disease detection), indoor air quality control, and structural health monitoring in civil engineering (e.g., detection of corrosive gases in enclosed infrastructures) [5]. These wide-ranging applications highlight the versatility of MOX-based sensing systems, reinforcing the need for robust physical and computational models capable of predicting their behavior under diverse and dynamic conditions [6–8].

Given the central role these sensors play in addressing both global challenges and advanced technological applications, there is a growing need to move beyond empirical calibration approaches and toward physically grounded, predictive frameworks. Despite their widespread adoption, the complex interplay of chemical reactions, material properties, and environmental factors that govern MOX sensor behavior remains only partially understood. Traditional experimental methods, while essential, are often insufficient to capture the shades of these multiscale, dynamic systems.

In this context, computational modeling emerges as a key enabler, offering a powerful toolset to simulate, analyze, and predict sensor behavior under varying operating conditions. Numerical models based on physical and chemical principles allow us to capture the essential mechanisms of gas–solid interaction, charge transfer, and transport phenomena in a controllable and interpretable way [9, 10]. These models can reproduce not only the steady-state behavior of sensors but also their dynamic response to transient stimuli such as temperature changes or varying gas mixtures.

More importantly, such models represent a crucial bridge between materials science and algorithmic intelligence. By providing an automatic simulation-ready description of sensor behavior, they enable the development of physics-informed algorithms for signal interpretation, anomaly detection, and predictive control. This is especially relevant in complex systems such as smart infrastructures or industrial plants, where integrating sensor models into higher-level decision-making pipelines can significantly improve robustness, adaptability, and interpretability.

Resistive gas sensors, exhibit a complex interplay between solid–gas surface chemistry and electronic conduction phenomena. These include adsorption, desorption, charge transfer, and interactions between different gas species, often taking place under dynamic thermal regimes [11]. Such processes are not easily accessible through direct measurement, making modeling an essential component in both research and application.

Accurate models must account for the structural and electronic properties of the sensing material, including its bulk and surface composition, the density and type of intrinsic and extrinsic defects, and the micro- and macrostructure of the film [12–14]. Surface-state models, which describe the formation of surface energy levels due to adsorbates, have proven particularly effective in explaining the conductance response for example in n-type materials.

Beyond their descriptive role, these models hold great potential as algorithmic components in broader intelligent systems. They can serve as interpretable surrogates for learning-based approaches, enable real-time simulation of sensor behavior in complex environments, and provide a physically grounded basis for predictive algorithms used in process monitoring, anomaly detection, or adaptive control of installations involving gas-sensitive elements. This places numerical modeling not only as a scientific tool, but also as a bridge between physics-based understanding and computational intelligence [15, 16].

In this paper, based on the authors findings in [13], we explore this computational modeling framework in the context of chemoresistive gas sensors, focusing on surface-reaction-driven behavior under variable operating conditions. By analyzing a dynamic model derived from physical-chemical principles, we highlight how such numerical tools can support both theoretical investigation and practical deployment in complex systems.

## 2 Chemical Basis of the Model: Surface Chemistry and Charge Mechanisms

### 2.1 Chemisorption and Surface Charge Formation

In resistive metal oxide sensors, the sensor response is mainly governed by surface reactions, specifically through gas adsorption and desorption that involve electron exchange (chemisorption). These surface interactions alter the sensor electrical conductivity. The electrical behavior of metal oxide sensors also depends on the preparation method, which influences the type and density of defects in the material and the microstructure. Surface states introduce donor or acceptor energy levels, they can be due to surface defects (intrinsic surface states), or to adsorbed gas molecules (extrinsic surface states) [17]. The capture or release of electrons by the surface states (in case of extrinsic state this coincides with adsorbates that ionize) affect the conductivity in two keyways: they generate/consume surface charges that create/modify surface electric fields and potential barriers, and they alter the concentration of free charge carriers. Depending on the energy levels of these states relative to the material band structure, the surface region can become either depleted or enriched in charge carriers. For example, in n-type oxides, oxidizing gases can lead to surface acceptor states that trap electrons, creating negative surface charge balanced by and a depleted region beneath with fixed positive charge to maintain electroneutrality.

For example, considering a porous film of n-type $SnO_2$ consisting of large grains, when oxidizing gases (such as $O_2$, $NO_2$) are present, they adsorb onto the grain surface and become ionized, forming negatively charged species (e.g., $O^-$, $NO_2^-$). The commonly accepted reaction path for oxygen (at temperature above 250 °C) includes a weak physisorption followed by dissociation and ionization [13]:

$$O + 2e^- + 2S \rightarrow 2O^-(ads) \quad (1)$$

where “S” denotes an active surface site. These charged species trap electrons from the conduction band, forming a surface depletion layer, with an associate surface voltage barrier, and thus increase the material resistance.

Conversely, reducing gases like CO or $H_2$ can react with pre-adsorbed oxygen species or adsorb directly, donating electrons to the conduction band and decreasing the voltage barrier height if present, reducing the resistance. The reaction involving pre-adsorbed oxygen is the following:

$$CO + O^{-}(ads) \rightarrow CO + e^{-} \tag{2}$$

Thus, extrinsic surface states are associated with adsorbed species, while the so named “intrinsic” surface states originate from structural defects or nonstoichiometry, such as oxygen vacancies. Both types of surface states contribute to the band bending at the material surface, creating potential barriers that influence charge carrier transport. Depending on the energy levels of these states relative to the conduction or valence bands, either depletion or accumulation layers can form, altering the sensor electrical characteristics. The effect is highly dependent on the size of the grain composing the oxide material, its morphology, and defect distribution, making the physical structure of the material a critical factor.

# 3 Modeling Approach: From Surface States to Resistance

## 3.1 From the Surface Charge to the Electrical Response (the Case of Large Grained Film)

To translate the complex chemistry involved in the sensing principle into an electrical signal, a physical model is required to connect the surface density of charged states, indicated as $N_S$, to the sensor electrical resistance.

As an example, the behavior of loosely bound large grained metal oxide film is discussed.

If the grain is large with respect to the depleted region depth, a one-dimensional model can be used and the surface potential $V_s$, due to trapped charges $N_s$ (negative for n-type material) can be written as:

$$V_s = \frac{1}{2\epsilon} q \frac{N_s^2}{n - p}, \tag{3}$$

where $q$ is the elementary charge, $\varepsilon$ is the dielectric permittivity of the metal oxide, n − p is the net bulk charge carrier density (for n-type, typically approximated as n). This expression is derived from the Poisson’s equation, describing how the electric potential varies within the depletion layer of a metal oxide grain due to surface charge [18, 19], under the assumption of complete depletion.

These quantities determine the conductance $G$ of the sensing material, which can be derived through thermionic emission or drift-diffusion models, and quantitatively described by the following equation [13]:

$$G = G_0 e^{(-qV_s/kT)}, \tag{4}$$

where $G_0$ is the baseline conductance, i.e., the conductance in the absence of a surface barrier (when $Vs = 0$), it depends on material parameters like carrier mobility and intrinsic carrier concentration, $k$ is the Boltzmann constant, and T the absolute temperature, which influences the thermal energy available to carriers. This relation shows an exponential sensitivity of conductance to the surface charge, providing a foundation for the response of the large grain films to gas concentration: even small changes in the surface potential $V_S$ (and therefore in the surface charge) can cause large variations in conductance. As the surface charge density increases (due to more adsorbed oxidizing gas), the barrier $V_S$ becomes higher, and the conductance $G$ decreases exponentially. Conversely, in the presence of reducing gases, the surface ions ($O^-$) are consumed, reducing the surface charge and the barrier $V_S$, which in turn leads to a significant increase in conductance. This behavior is the core sensing principle of $SnO_2$ and similar n-type oxide sensors: the electrical resistance of the material (or its inverse, conductance) changes in response to the surrounding gas composition, driven by changes in surface charge and potential barrier height.

Note that if the film morphology is different, the Poisson's equation is solved under different boundary conditions and (3) must be recalculated. In particular Eqs. (3) and (4) must be changed when the grainsize becomes comparable with the depleted region thickness (nanograin case). The relationship between surface charge and resistance will be discussed in more detail in the next section.

### 3.2 Film- Conductance and Dependence on the Microstructure

The sensing film layer realizing the sensor is typically composed of many grains (nano- and micro sized). Depending on the microstructure of the material (porous vs compact), different conduction pathways dominate into the obtained layer:

- In porous films: resistance is dominated by inter-grain barriers.
- In compact films: bulk resistance and parallel surface paths are also significant.

Models must consider whether each grain is fully or partially depleted and the ratio between grain size D, and Debye length $L_D$, which indicates the thickness of the depleted region, must be considered since it critically affects conduction. When D is comparable to $2L_D$, grains are close to the full depletion, and conductance is highly sensitive to surface charge variations, until saturation is reached. Larger grains ($D \gg 2L_D$) retain a conductive core, reducing sensitivity to gas-induced surface effects.

### 3.3 Gas Reaction Kinetics

To model dynamic behavior and the surface charge evolution must be linked to chemical kinetics. The goal is to relate dynamic gas concentration in the environment to the trapped charge density in time $N_s(t)$, which in turn modulates the conductance via the surface potential barrier $V_S$. As previously discussed, assuming an n-type material, the presence of oxidizing gas, which captures electrons from the conduction band, leads to an increase in $N_S$. In contrast, reducing gas, which reacts with ionized adsorbed oxygen and releases electrons, causes a decrease in $N_S$.

**Dynamic Reaction Kinetics.** The reaction process can be described quantitatively with a set of differential equations that govern the time-dependent surface coverage of adsorbed species and the resulting surface charge $N_s(t)$. In the following treatment we consider a n-type material, considering as an example the $SnO_2$.

The adsorption process, by following the Wolkenstein theory [20], is modeled as a two-step reaction involving the physisorption reaction of neutral gas molecules onto surface sites and ionization reactions of the adsorbates, involving electron transfer from or to the semiconductor [21, 22].

If the film is in air, in the presence of oxygen, the reactions are:

$$O_2 + S \underset{k_1^-}{\overset{k_1^+}{\rightleftharpoons}} O_2(ads) \tag{5}$$

$$O_2(ads) + e^- \underset{k_2^-}{\overset{k_2^+}{\rightleftharpoons}} 2O_-(ads) \tag{6}$$

describing respectively, the weak physisorption, Eq. (5) and ionization, Eq. (6), where each of the reactions has its own reaction rate constant ($\overset{k_{1,2}^+}{\rightarrow}$) and backward reaction rate constant ($\underset{k_{1,2}^-}{\leftarrow}$).

These reactions can be translated into the following system of ordinary differential equations (ODEs) [13]:

$$\frac{d\theta_1}{dt} = k_1^+ cO_2(1 - \theta_1) - k_1^- \theta_1 - k_2^+ \theta_1 n + k_2^- \theta_2 \tag{7}$$

$$\frac{d\theta_2}{dt} = 2k_2^+ \theta_1 n - 2k_2^- \theta_2 \tag{8}$$

where $\theta_1$ and $\theta_2$ indicate respectively the coverage (fraction of occupied surface sites) of weakly adsorbed and ionized oxygen species, which is related to the $N_s$ parameter ($N_s \propto \theta_2$), $n$ is the free electron concentration and $cO_2$ the oxygen concentration in gas phase.

In contrast, reducing gases (e.g., CO, $H_2$, $CH_4$) interact with the ionized oxygen species already present on the surface following Eq. (2) in which CO has been assumed as the reducing gas. These gases donate electrons by reacting with the negatively charged oxygen, thereby removing surface charge and releasing electrons back into the conduction band. This process consumes the charged oxygen species, effectively reducing $\theta_2$, and consequently decreasing the surface charge density $N_s$. To account for the effect of reducing gases in the dynamic equation for $\theta_2$ (ionized oxygen coverage), and denoting $k_r$ as the rate constant of the reaction between the reducing gas and surface oxygen, Eq. (8) becomes:

$$\frac{d\theta_2}{dt} = 2k_2^+ \theta_1 n - 2k_2^- \theta_2 - k_r c_r \theta_2 \tag{9}$$

where $c_r$ indicate reducing gas concentration in atmosphere.

Thus, as $\theta_1$, $\theta_2$ and consequently $N_s$ changes due to reaction with gas species, the surface potential Vs also changes. Thus, considering the reactions kinetics, for an n-type material with a fully depleted region Eq. (3) becomes [13]:

$$V_s(t) = \frac{qN_s^2(t)}{2\epsilon n} \tag{10}$$

Accordingly, the final expression of the dynamic conductance of the sensing film can be written as:

$$G(t) = G_0 \exp^{\left(-\frac{q^2 N_s^2(t)}{2\epsilon nkT}\right)} \tag{11}$$

This expression provides a direct, time-dependent relationship between gas-surface interactions and the measurable electrical signal of the sensor, enabling time-resolved simulation of the sensor response under transient gas exposures and temperature shifts.

## 4 Results and Experimental Validation

### 4.1 Application to $SnO_2$-Based Sensor

The described model has been validated using experimental data from commercial sensors (in this case the Figaro TGS2600 a device based on $SnO_2$). The sensors were mounted in a controlled test chamber where gas flow, temperature, and humidity could be precisely regulated [13][22]. To evaluate the accuracy and predictive capabilities of the proposed surface-state-based dynamic model, experimental measurements were conducted under controlled exposures to different concentrations of $O_2$ and CO. These measurements serve not only to validate the qualitative behavior predicted by the model but also to enable a quantitative comparison by fitting key parameters.

The system of nonlinear ordinary differential equations described in Eqs. (7)–(9) was implemented and solved in MATLAB using adaptive time–stepping solvers. To calibrate the model parameters, including adsorption/desorption rate constants and surface-site densities, a nonlinear least-squares fitting procedure based on the Levenberg–Marquardt algorithm was applied. The fitting was performed using the experimental time-resolved conductance data under controlled gas exposures, minimizing the root-mean-square error between simulated and measured signals. This method ensures both convergence stability and robustness against parameter correlation, as demonstrated in prior sensor modeling works [18, 22]. The obtained parameters were subsequently used to simulate the transient and steady-state behavior of the sensor under different gas and temperature conditions, enabling direct comparison with experimental results.

In particular, the time-resolved conductance data collected during exposure to oxidizing and reducing gases were fitted using the model equations. This approach allowed for the estimation of material-specific parameters, such as reaction rate constants and surface site densities, which are otherwise difficult to obtain purely from theory.

To initialize the dynamic surface-state model, baseline calibration was carried out using conductance measurements in a dry $N_2$ environment. This inert condition prevents active gas–surface reactions, allowing the intrinsic electrical properties of the sensing materials to be isolated.

In this configuration, the conductance response in dry $N_2$ was used to estimate two key model parameters: the intrinsic conductance $G_0$, representing the sensor conductance in the absence of surface potential barriers, and the baseline surface charge density $N_s$, associated with trapped charges due to pre-existing defects. These values were obtained through a fitting procedure in which the model output was aligned with the experimental time-resolved conductance data. In Fig. 1 the results of the experimental and simulated response of the sensors in dry $N_2$ with different temperature profiles ranging between 300 °C and 400 °C are reported, the derived value of $G_0$ for the reported case is $G_0 =$ 30000 S. The figure shows the derived Ns′ parameters that is a is a scaled version of the surface charge Ns, and, in particular it has been normalized to group physical constants into a simpler form, that simplifies temperature-dependent exponential relationships: $N\prime_s(t) = N_s(t)/\sqrt{(2\epsilon nkq)}$.

The fitting results show excellent agreement between model predictions and measurements, confirming the validity of the approach. The extracted parameters serve as the initial conditions for simulating the sensor response under exposure to oxidizing and reducing gases in subsequent tests.

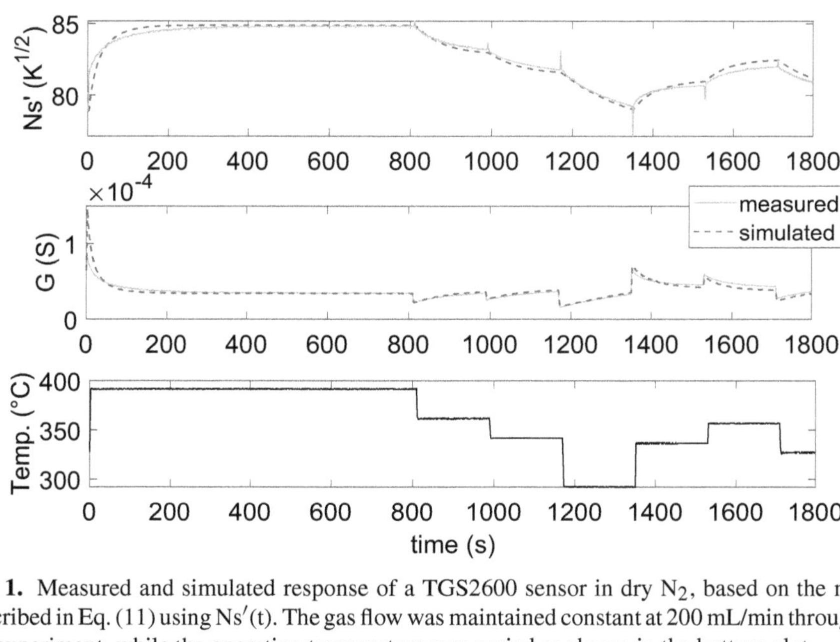


**Fig. 1.** Measured and simulated response of a TGS2600 sensor in dry $N_2$, based on the model described in Eq. (11) using Ns′(t). The gas flow was maintained constant at 200 mL/min throughout the experiment, while the operating temperature was varied as shown in the bottom plot.

Following the model initialization in inert atmosphere (Fig. 1), the dynamic response of the sensor was evaluated under exposure to dry CO in a concentration of 100 ppm. This condition isolates the interaction between the reducing gas and the metal oxide surface (direct chemisorption with no involvement of preadsorbed oxygen). The test was conducted between 300 °C and 400 °C, a temperature that ensures appropriate surface reactivity while maintaining sensor stability. Also in this case, the experimental data collected during this CO exposure were fitted using the dynamic model.

Under these conditions, CO molecules react directly with the metal oxide surface, and an appropriate ODE describing the reaction kinetics was exploited. This interaction releases electrons into the conduction band, resulting in a decrease in surface charge density $N_S(t)$, as can be observed comparing data in Fig. 1 and Fig. 2. Accordingly, a lowering of the surface potential barrier $V_S(t)$ will occur and as a consequence, the model predicts an increase in conductance G(t).

Results in Fig. 2, show that the simulated conductance closely matches the measured response over time. The model accurately captures both the transient behavior and the steady-state level, confirming its ability to describe the dynamic electron exchange processes induced by a reducing gas. The slight mismatch in rise time may reflect kinetic limitations at the surface, such as slower reaction rates or desorption dynamics, but overall, the agreement is strong.

A similar accuracy was found when modeling the response to CO in air (in the presence of oxygen), when also the reactions with ionize preadsorbed oxygen are modeled.

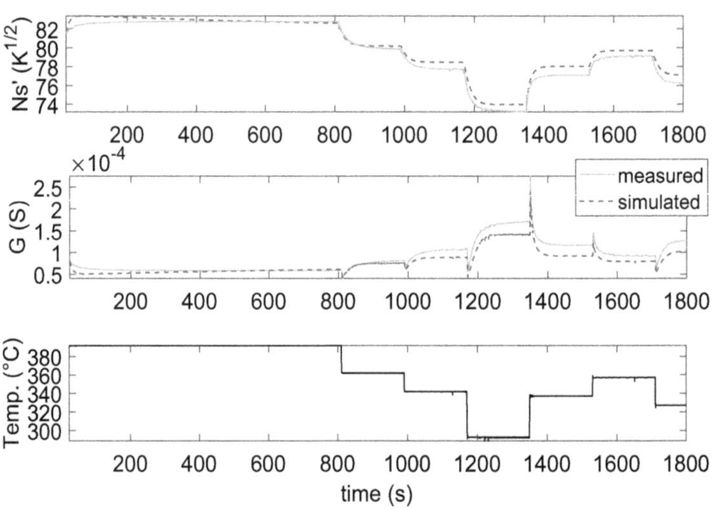


**Fig. 2.** Measured and simulated response of a TGS2600 sensor in dry $N_2$ and 100 ppm of CO, based on the model described in Eq. (11) using $Ns'(t)$. The gas flow was maintained constant at 200 mL/min throughout the experiment, while the operating temperature was varied as shown in the bottom plot.

Model predictions align well with measured resistance curves, capturing key features such as time constants of sensor response and recovery, effects of varying CO concentrations, and the influence of operating temperature.

This level of agreement supports model validity and its usefulness in sensor performance prediction. Although demonstrated here using only a limited gas set ($N_2$, CO, and

$O_2$) and a specific material system, the model is inherently versatile. It can be extended to other reducing or oxidizing gases and adapted to different metal oxide materials by incorporating appropriate kinetic parameters and adsorption mechanisms.

**Conclusion.** Beyond matching experimental data, the model provides a mechanistic basis for algorithmic development. Its parameters can be tuned to reflect changes in environmental conditions or sensor aging, enabling a real-time calibration algorithm, fault detection in sensor arrays and the integration into AI-driven monitoring and control systems.

Thus, this surface-state-based framework is not only a descriptive tool but also a foundational block for intelligent gas-sensing architectures in complex environments.

Although this work focuses on the physical modeling of chemoresistive sensors, the developed framework can be integrated into data-driven or bio-inspired learning systems. The numerical model can serve as a physics-informed layer for adaptive algorithms that mimic evolutionary optimization or self-organization processes, enabling online calibration or fault detection in large sensor networks [15, 16].

Future work will explore coupling the proposed model with evolutionary optimization and machine-learning approaches for parameter identification and sensor behavior prediction in complex environments.

**Disclosure of Interests.** The authors have no competing interests to declare that are relevant to the content of this article.

## References

1. Staerz, A., Weimar, U., Barsan, N.: Current state of knowledge on the metal oxide based gas sensing mechanism. Sens. Actuat. B: Chem. **358**, 131531 (2022). ISSN 0925-4005
2. Güntner, A.T., Pineau, N.J., Pratsinis, S.E.: Flame-made chemoresistive gas sensors and devices. Prog. Energy Combust. Sci. **90**, 100992 (2022)
3. Taha, S.S., et al.: Comprehensive review of health impacts of the exposure to nitrogen oxides (NOx), carbon dioxide (CO2), and particulate matter (PM). J. Hazardous Mater. Adv. **19**, 100771 (2025)
4. Kabir, M., Habiba, U.E., Khan, W., et al.: Climate change due to increasing concentration of carbon dioxide and its impacts on environment in 21st century; a mini review. J. King Saud Univ. - Sci. **35**(5), 102693 (2023). ISSN 1018-3647
5. Uma, S., Shobana, M.K.: Metal oxide semiconductor gas sensors in clinical diagnosis and environmental monitoring. Sens. Actuat. A: Phys. **349**, 114044 (2023). ISSN 0924-4247
6. Fort, A., Met al.: Modeling the conductivity response to NO2 gas of films based on MWCNT networks. Sensors **21**(14), art. no. 4723 (2021)
7. Guarino, R., et al.: Modelling electronic transport in monocrystalline metal oxide gas sensors: from the surface kinetics to the experimental response. Sens. Actuat. B Chem. **373**, 132646 (2022)
8. Fort, A., et al.: Co3 O4/Al-ZnO nano-composites: gas sensing properties. Sensors **19**(4), art. no. 760 (2019)
9. Kumar, A., Kumar, M., Kumar, R., Singh, R., Prasad, B., Kumar, D.: Numerical modelling of Chemisorption of oxygen gas molecules on the surface of semiconductor for gas sensors applications. Mater. Today: Proc. **18**(Part 3), 1272–1279 (2019)

10. Fort, A., Trigona, C., Panzardi, E., Vignoli, V., Addabbo, T., Mugnaini, M.: An AlN micromachined mass sensor: modeling and characterization. IEEE Trans. Instrum. Meas. **70**, art. no. 9144611 (2021)
11. Gherardi, S., et al.: Investigating the temperature-dependent kinetics in humidity-resilient tin–titanium-based metal oxide gas sensors. Chemosensors **12**, 151 (2024)
12. Porte, Y., Maller, R., Faber, H., Alshareef, H., et al.: Exploring and controlling intrinsic defect formation in SnO2 thin films. J. Mater. Chem. C. **4** (2015). https://doi.org/10.1039/C5TC03520A
13. Fort, A., et al.: Surface state models for conductance response of metal oxide gas sensors during thermal transients. In: Chemical Sensors: Simulation and Modeling Volume 2: Conductometric-Type Sensors, vol. 2, pp. 127–175. Momentum Press (2012)
14. Laidani, N., et al.: Intrinsic defects and their influence on the chemical and optical properties of TiO2-x films. J. Phys. D Appl. Phys. **43**, 485402 (2010)
15. Qiuchen, Y., et l.: A generative artificial intelligence model for efficient gas sensitivity prediction in materials without parameters from first principle calculation. Sens. Actuat. A: Phys. **391**, 116636 (2025). ISSN 0924-4247
16. Zhou, G., Du, B., Zhong, J., Chen, L., Sun, Y., et al.: Advances in gas detection of pattern recognition algorithms for chemiresistive gas sensor. Materials **17**, 5190 (2024)
17. Dow, J.D., Allen, R.E., Sankey, O.F.: Intrinsic and extrinsic surface electronic states of semiconductors. In: Chemistry and Physics of Solid Surfaces V. Springer, Heidelberg, pp. 483–500 (1984)
18. Fort, A., Rocchi, S., Serrano-Santos, M.B., Spinicci, R., Vignoli, V.: Surface state model for conductance responses during thermal-modulation of sno2-based thick film sensors: part I – model derivation'. IEEE Trans. Instrum. Meas. **55**(6), 2102–2106 (2006)
19. Gomri, S., Seguin, J.-L., Guerin, J., Aguir, K.: Adsorption–desorption noise in gas sensors: modelling using Langmuir and Wolkenstein models for adsorption. Sens. Actuat. B: Chem. **114**(1), 451–459 (2006)
20. Wolkenstein, T.: Electronic Properties on Semiconductor Surface during Chemisorption. Consultants Bureau, New York (1991)
21. Barsan, N., Koziej, D., Weimar, U.: Metal oxide-based gas sensor research: how to? Sens. Actuat. B Chem. **121**(1), 18–35 (2007)
22. Barsan, N., Weimar, U.: Conduction model of metal oxide gas sensors. J. Electroceram. **7**, 143–167 (2001). https://doi.org/10.1023/A:1014405811371

# Author Index





M. Bianchini et al. (Eds.): WIVACE 2025, CCIS 3001, pp. 197–198, 2027.
https://doi.org/10.1007/978-3-032-33185-4

GPSR Compliance
The European Union's (EU) General Product Safety Regulation (GPSR) is a set of rules that requires consumer products to be safe and our obligations to ensure this.

If you have any concerns about our products, you can contact us on

ProductSafety@springernature.com

In case Publisher is established outside the EU, the EU authorized representative is:

Springer Nature Customer Service Center GmbH
Europaplatz 3
69115 Heidelberg, Germany

www.ingramcontent.com/pod-product-compliance
Ingram Content Group UK Ltd.
Pitfield, Milton Keynes, MK11 3LW, UK
UKHW021642180926
475650UK00002B/89

* 9 7 8 3 0 3 2 3 3 1 8 4 7 *